**Bentham E-book series**
**Digital Signal Processing in Experimental Research**
L. Yaroslavsky and J. Astola, Editors

**Volume 1**

Leonid Yaroslavsky

# Introduction to Digital Holography

# Table of Contents

# Foreword

## Survival of Holography provided by Signal Processing

Natural scientists belong to one of two categories: either they deal with matter and energy, or they are concerned with signals and systems. Sometimes collaboration of these two kinds of scientists is essential for solving problems at hand. The emergence of holography was a typical case, where systems people saved the physics people. Originally holograms caused amusing displays. Now holograms are also used at industrial jobs. To do that the physical optics people had to adapt the working style of systems people to what is now often called "digital holography", the subject of this book.

Back in 1948 Dennis Gabor succeeded in producing holograms. The objects usually consisted of a few dark lines on a large transparent background. The method did not work properly with a negative as object, i.e. a few bright narrow lines. More annoying was the twin image that appeared at an undesirable location. Gabor attempted to solve this problem with the help of tools from physical optics: lenses, prisms, beam-splitters in a Mach-Zehnder interferometer. No luck, he gave up. When appointed Professor at the Imperial College, London, in 1959, he presented, in his acceptance speech, his three most noticeable inventions. Holography was not a part of it.

Meanwhile, I had learned from a systems scientist what single side-band modulation can do. We tried it with a modest success, shortly before Emmett Leith solved the problems with his tilted reference method. Leith, in his memoirs, wrote that he was lucky to be a physicist surrounded by signal-processing experts. Systems theory is much richer in concepts beyond side-band processing. Another interesting concept is matched filtering. It was soon translated from electronic filtering to optical spatial filtering.

This new book is a tremendous collection of digital signal and image processing in general, and specifically on digital holography. It presents digital holography in one unified language called "algorithmics". The volume appears at an appropriate time, i.e. while optics is mutating from micro to nano with typical dimensions $10^{-3}$ and $10^{-6}$ microns.

I am amazed when I compare today's capabilities with what was available around 1950-1960. When will we have mobile-3D-in color impressions? The youth is so powerful. They will simply demand it, and get it because of the market. Good luck!

Prof. Dr. A. Lohmann
University Erlangen-Nuernberg,
Germany

# Preface

The history of science is, to a considerable degree, the history of invention, development and perfecting of imaging methods and devices. Modern science began with the invention and application of optical telescope and microscope in the beginning of 17-th century by Galileo Galilei (1564-1642), Anthony Leeuwenhoek (1632-1723) and Robert Hook (1635-1703).

The next decisive stage was invention of photography in the first half of the 19-th century. Photographic plates were the major means in discoveries of X-rays by Wilhelm Conrad Roentgen (1845-1923, The Nobel Prize Laureate, 1901) and radioactivity by Antoine Henri Becquerel (1852-1908, The Nobel Prize Laureate, 1903) at the end of 19-th century. These discoveries, in their turn, almost immediately gave birth to new imaging techniques: to X-ray imaging and radiography.

X-rays were discovered by W.C. Roentgen in experiments with cathode rays. Cathode rays were discovered in 1859 by Julius Plücker (1801-1868), who used vacuum tubes invented in 1855 by a German inventor Heinrich Geissler (1815-1879). These tubes, as modified by Sir William Crookes (1832-1919) led eventually to the discovery of the electron and finally brought about the development of electronic television and electron microscopy in 30-th - 40-th of 20-th century. Designer of the first electron microscope E. Ruska (1906-1988) was awarded The Nobel Prize in Physics for 1986. This award was shared with G. Binning and H. Rohrer, who were awarded for their invention of the scanning tunneling microscope.

The discovery of diffraction of X-rays by Max von Laue (1889-1960, the Nobel Prize Laureate, 1914) in the beginning of 20-th century marked the advent of a new imaging technique, which we now call computational imaging. Although Von Laue's motivation was not creating a new imaging technique but rather proving the wave nature of X-rays, lauegrams had very soon become the main imaging tool in crystallography. Lauegrams are not conventional images for visual observation. However, using "lauegrams", one could numerically reconstruct the spatial structure of atoms in crystals. This shoot gave its crop in less than half a century. One of the most remarkable scientific achievements of 20-th century is based on X-ray crystallography. It is discovery by J. Watson and F. Crick of spiral structure of DNA (the Nobel Prize, 1953). And about at the same time a whole bunch of new computational imaging methods had appeared: holography, synthetic aperture radar, coded aperture imaging, tomography. Two of these inventions were awarded the Nobel Prize: D. Gabor for "his invention and developments of the holographic method" (the Nobel Prize in Physics, 1971) and A. M. Cormack and G. N. Hounsfield for "the development of computer assisted tomography" (the Nobel Prize in Physiology and Medicine, 1979).

Denis Gabor invented holography in 1948. This is what D. Gabor wrote in his Nobel Lecture about the development of holography: " Around 1955 holography went into a long hibernation. The revival came suddenly and explosively in 1963, with the publication of first successful laser hologram by Emmett N. Leith and Juris Upatnieks of the University of Michigan, An Arbor. Their success was due not only to the laser, but to the long theoretical preparation of Emmett Leith (in the field of the "side looking radar") which started in 1955. Another important development in holography [happened] in 1962, just before the "holography explosion". Russian physicist Yu. N. Denisyuk published an important paper in which he combined holography with the ingenious method of photography in natural colors, for which Gabriel Lippman received the Nobel Prize in 1908".

Denis Gabor received Nobel Prize in 1971. The same year, a paper "Digital holography" was published in Proceedings of IEEE by T. Huang. This paper marked the next step in the development of holography, the use of digital computers for reconstructing, generating and simulating wave fields, and reviewed pioneer accomplishments in this field. These accomplishments prompted a burst of research and publications in early and mid 70-th. At that time, most of the main ideas of digital holography were suggested and tested. Numerous potential applications of digital holography such as fabricating computer-generated diffractive optical elements and spatial filters for optical information processing, 3-D holographic displays and holographic television and computer vision stimulated a great enthusiasm among researchers.

However, limited speed and memory capacity of computers available at that time, absence of electronic means and media for sensing and recording optical holograms hampered implementation of these potentials. In 1980-th digital holography went into a sort of hibernation similarly to what happened to holography in 1950-th - 1960-th. With an advent, in the end of 1990-th, of the new generation of high speed microprocessors, high resolution electronic optical sensors and liquid crystal displays, of a technology for fabricating micro lens and mirror arrays digital holography is getting a new wind. Digital holography tasks that required hours and days of computer time in 1970-th can now be solved in almost "real" time for tiny fractions of seconds. Optical holograms can now be directly sensed by high-resolution photo electronic sensors and fed into computers in "real" time with no need for any wet photo-chemical processing. Micro lens and mirror arrays promise a breakthrough in the means for recording computer-generated holograms and creating holographic displays. Recent flow of publications in digital holographic metrology and microscopy indicate revival of digital holography from the hibernation§.

The development of optical holography, one of the most remarkable inventions of the XX-th century, was driven by clear understanding of information nature of optics and holography. The information nature of optics and holography is especially distinctly seen in digital holography. Wave field recorded in the form of a hologram in optical, radio frequency or acoustic holography, is represented in digital holography by a digital signal that carries the wave field information deprived of its physical casing. With digital holography and with incorporating digital computers into optical information systems, information optics has reached its maturity.

This is not a coincidence that digital holography appeared in the end of 60-th, the same period of time, which digital image processing can be dated back to. In the same way, in a certain sense, as inventing by Ch. Towns, G. Basov and A. Prokhorov lasers in mid 1950-th (the Nobel Prize in Physics, 1964) stimulated development of holography, two events stimulated digital holography and digital image processing: beginning of industrial production of computers in 1960-th and introducing Fast Fourier Transform algorithm made by J. W. Cooley and J. M. Tukey in 1965.

In this book, we will adhere to meaning of digital holography as of a branch of the imaging science, which deals with numerical reconstruction of digitally recorded holograms and with computer synthesis of holograms and diffractive optical elements.

We start, in Ch. 1, with basic principles of physical holography and its mathematical models and introduce diffraction integrals that are used to describe wave propagation from objects to holograms. The fundamental issue of digital holography is discrete representation of optical signals and diffraction integrals. This problem, which has a direct relation to both numerical reconstruction of holograms and to synthesis of computer generated holograms, is addressed in Ch. 2. Then we proceed to methods of digital holography proper.

Methods and algorithms for digital recording and numerical reconstruction of holograms, their applicability and appropriate metrological characterization are presented in Ch. 3.

Chs. 4 through 6 are devoted to principles of computer-generated holography and mathematical models, to methods of encoding numerical holograms for recording them on spatial light modulators as hologram recording devices and to the analysis of how the results of optical reconstruction of computer-generated holograms depend on the encoding method and on physical parameters of hologram recording devices.

Chs. 7 and 8 review applications of computer-generated holograms in optical information processing and for information display and 3D visual communications. In the latter, a especial emphasis is made on using limitations of human 3D vision for reducing the computational complexity of the hologram synthesis and for easing requirements to hologram recording and reconstruction devices.

In its methods and applications, digital holography is closely connected with digital image processing. Digital image processing is nowadays a well-established field of information technology covered in many books and educational courses. However some of its aspects that have direct relation to digital holography and its applications deserve discussion and reviewing in the context of the book on digital holography. Therefore, in Ch. 9 we review image processing methods in digital holography: mathematical models of imaging and holographic systems; statistical models of stochastic transformations of optical signals; measuring parameters of random interferences in sensors and imaging and holographic systems; principles of Mean Square Error (MSE)-optimal scalar Wiener filtering for image denoising and deblurring; methods for correcting image gray scale nonlinear distortions; methods for accurate image resampling.

# 1    Principles of Holography

**Abstract**: In this chapter, we outline physical principles of holography and introduce mathematical models of recording and reconstruction of hologram and diffraction integrals used in these models.

## 1.1    Invention of holography

The term "holography" originates from Greece word "holos" (ηωλωσ). By this, inventor of holography Dennis Gabor intended to emphasize that in holography full information regarding light wave, both amplitude and phase, is recorded. This is achieved by means of recording interference of two beams of coherent light, object and reference ones ([1]).

Gabor carried out model optical experiments to demonstrate the feasibility of the method (Fig. 1.1). However, powerful sources of coherent light were not available at the time, and holography remained an "optical paradox" until the invention of lasers.

**Fig. 1.1.** Holographic portrait of Dennis Gabor (left), the first hologram (right, upper image), original \(right, bottom left image) and reconstructed (right, bottom right image) images

The very first implementations of holography were demonstrated in 1961-62 by radioengineers E. Leith and J. Upatnieks in Michigan University ([2]) and by a physicist Yu. Denisyuk in the State Optical Institute, St. Petersburg, Russia ([3]).

E. Leith and J. Upatnieks suggested the "off-axis" technique for fabricating display holograms by means of recording on a photographic plate interference pattern produced by coherent light beam reflected from the object and by the reference beam reflected from a mirror as it is shown on a schematic diagram in  Fig. 1.2, a). They also demonstrated that when the hologram is illuminated by the same reference beam and viewed with naked eye, it reconstructs 3-D image of the recorded object.

Yu. Denisyuk suggested recording display holograms as an interference pattern formed on a photographic plate by object and reference beams coming from opposite directions  (Fig. 1.3, a). He demonstrated that if the photographic plate emulsion is "thick" enough to be able to record in its volume a 3-D interference pattern of object and reference beams, one does not necessarily need a source of coherent light for hologram reconstruction and the hologram, being illuminated by white light, reconstructs in front of itself visual image of the object (Fig. 1.3, b)).

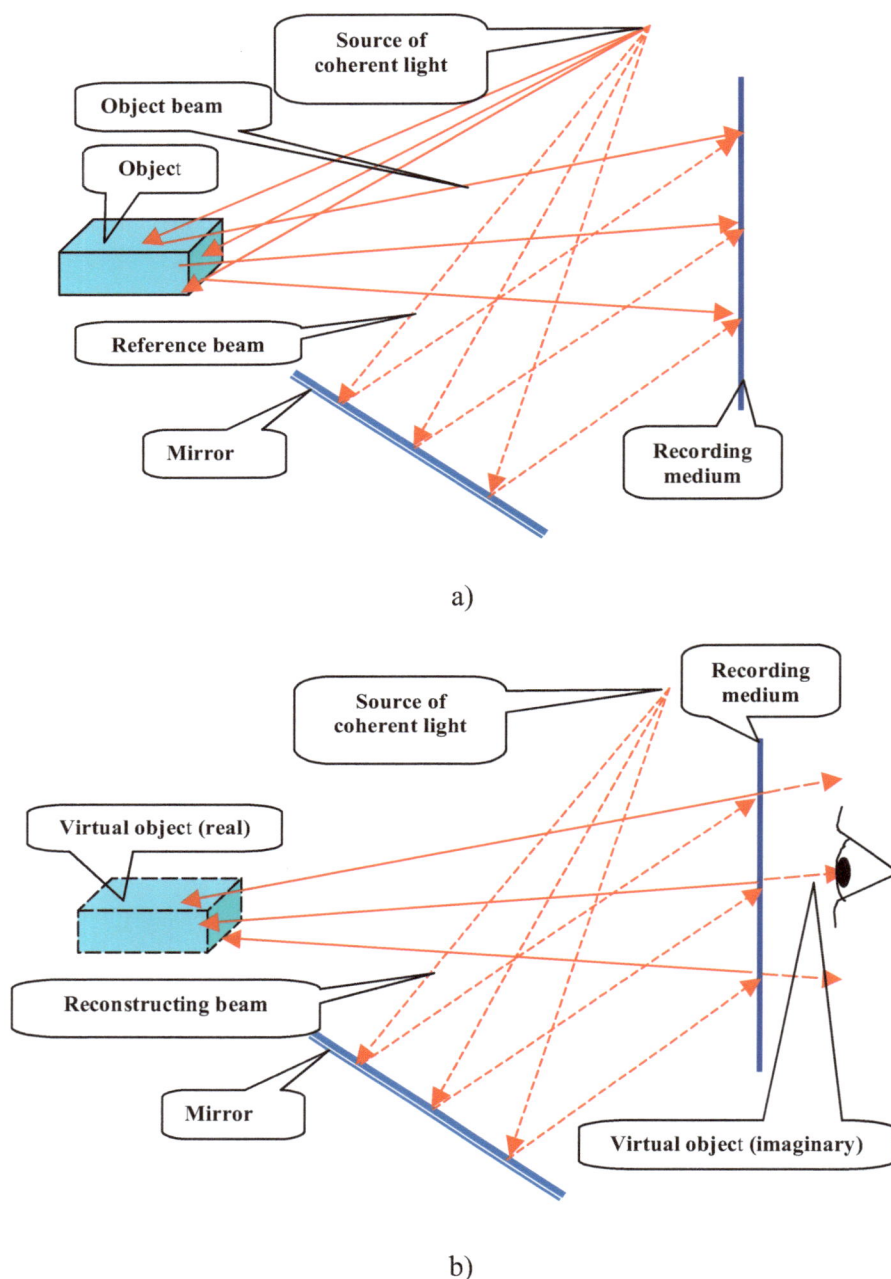

a)

b)

**Fig. 1.2.** Schematic diagram of Leith-Upatnieks's method for recording and reconstruction of transmittance holograms: a) recording hologram; b) hologram play back

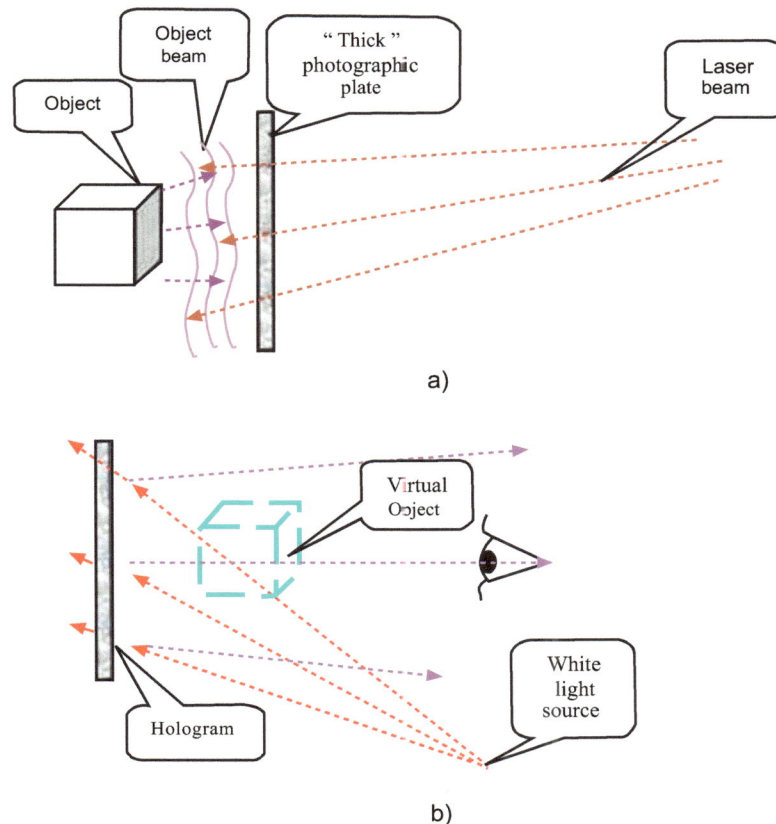

**Fig. 1.3.** Schematic diagram of Denisyuk's volume hologram recording (a) and of hologram play back (b)

While Leith-Upatnieks off-axis hologram recording scheme has found numerous applications in optical metrology and optical information processing, Denisyuk's method of recording volume holograms prevails in fabrication of high quality display holograms.

Digital holography dates back to late 1960- early 1970-th. Its development was motivated by the desire to take advantage of flexibility of digital computers for producing numerically designed complex valued spatial filters for optical information processing, by the dream of 3-D information display, for which holography promised an ultimate solution, and by the intention to use computers for numerical analysis of information carried by optical holograms.

Thus, two complementing tasks of digital holography are:
- Synthesis of computer-generated holograms from a mathematical model or a digital signal and recording the result of the synthesis on a physical media for fabrication of a physical hologram that can work in optical setups as an optical element or viewed by naked eye if the hologram is synthesized for display purposes
- Recording physical holograms directly in a digital form suited for input to a computer and using recorded digital signal for numerical reconstruction of the

hologram to obtain object wave front amplitude and phase spatial distribution for its subsequent numerical analysis

## 1.2   Mathematical models of recording and reconstruction of holograms

Consider a general schematic diagram of recording holograms shown in Fig. 1.4. In hologram recording, an object, whose hologram is to be recorded, is illuminated by a coherent radiation from a radiation source, which simultaneously illuminates also a hologram-recording medium with a "reference" beam. Usually, the photosensitive surface of the recording medium is a plane. We refer to this plane as a hologram plane. At each point of the hologram plane, the recording medium, whether it is continuous, such as a photographic film, or discrete, such as a photosensitive array of CMOS or CCD digital cameras, measures energy of the sum of the reference beam and the object beam reflected or transmitted by the object.

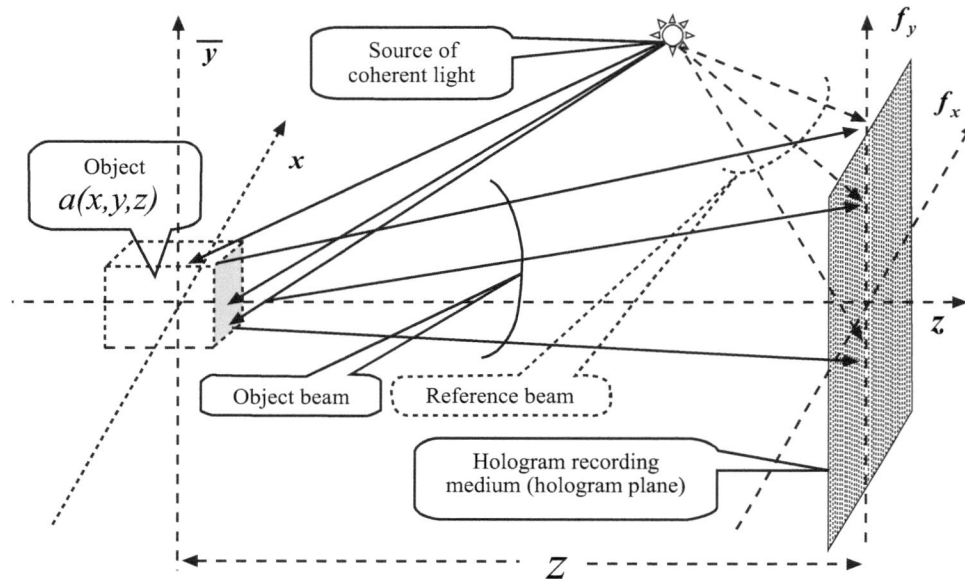

**Fig. 1.4.** Schematic diagram of recording holograms

Conventional mathematical models of recording and reconstruction of holograms assume that:

(i)   monochromatic coherent radiation that can be described by its complex amplitude as a function of spatial coordinates is used for hologram recording and reconstruction and

(ii)   object characteristics defining its ability to reflect or transmit incident radiation are described by radiation reflection or transmission factors, which are also functions of spatial coordinates. Specifically, if $I(x,y,z)$ is complex amplitude of the object illumination radiation at point $(x,y,z)$, complex amplitude $a(x,y,z)$ of the radiation reflected or transmitted by the object at this point is defined by the equation:

$$a(x,y,z) = I(x,y,z)O(x,y,z) \qquad\qquad (1.2.1)$$

where $O(x,y,z)$ is the object reflection or, correspondingly, transmission factor.

Light sensitive detectors produce signal, which is proportional to the energy of the light. Let $\alpha(f_x,f_y)$ and $R(f_x,f_y)$ denote complex amplitudes of the object and reference beams, respectively, at point $(f_x,f_y)$ of the hologram plane. Then signal recorded by the recording medium at this point is a squared module of their sum:

$$\mathrm{H}(f_x,f_y) = \left| \alpha(f_x,f_y) + R(f_x,f_y) \right|^2 =$$
$$\alpha(f_x,f_y)R^*(f_x,f_y) + \alpha^*(f_x,f_y)R(f_x,f_y) + \left|\alpha(f_x,f_y)\right|^2 + \left|R(f_x,f_y)\right|^2, \quad (1.2.2)$$

where asterisk denotes complex conjugate. This signal is a hologram signal, or a hologram. The first term in the sum in the right hand part of Eq. 1.2.2 is proportional to the object's beam complex amplitude. We will call this term a "***mathematical hologram***". Hologram reconstruction consists in applying to the mathematical hologram a transform that implements wave back propagation from the hologram plane to the object. For this, one has either to eliminate, before the reconstruction, other three terms or to apply the reconstruction transform to the entire hologram and than separate the contribution of other terms in the reconstruction result from that of the mathematical hologram term.

As it is known, wave propagation transformations are, from the signal theory point of view, linear transformations, which satisfy the superposition principle. As such, they are mathematically modeled as integral transformation. Thus, object complex amplitude of radiation $a(x,y,z)$ and object beam wave front $\alpha(f_x,f_y)$ at hologram plane are related through "forward propagation"

$$\alpha(f_x,f_y) = \int\limits_{-\infty}^{\infty}\int\limits_{-\infty}^{\infty}\int\limits_{-\infty}^{\infty} a(x,y,z)WPK(x,y,z;f_x,f_y)\,dx\,dy\,dz \qquad (1.2.3,\,a)$$

and "backward propagation"

$$a(x,y,z) = \int\limits_{-\infty}^{\infty}\int\limits_{-\infty}^{\infty} \alpha(f_x,f_y)WPK(x,y,-z;f_x,f_y)\,df_x\,df_y \qquad (1.2.3,\,b)$$

integral transforms, where $WPK(.;.)$ is a transform (wave propagation) kernel.

Commonly, this model is simplified by considering wave propagation between plane slices of the object and the hologram plane:

$$\alpha(f_x,f_y) = \int\limits_{-\infty}^{\infty}\int\limits_{-\infty}^{\infty} a(x,y,0)WPK(x,y,Z;f_x,f_y)\,dx\,dy \qquad (1.2.4,\,a)$$

and

$$a(x, y, 0) = \int\limits_{-\infty}^{\infty} \int\limits_{-\infty}^{\infty} \alpha(f_x, f_y) WPK(x, y, -Z; f_x, f_y) df_x \, df_y \,, \qquad (1.2.4, \text{b})$$

where $Z$ is the distance between the object and hologram planes. Equations (1.2.4) are called diffraction transforms (integrals).

## 1.3 Diffraction integrals

Light propagation from object to hologram is treated in digital holography with the help of the scalar diffraction theory. According to the theory, wave front $\alpha(\mathbf{f})$ of an object $a(\mathbf{x})$ is defined by the Kirchhoff-Rayleigh-Sommerfeld integral ([4]):

$$\alpha(\mathbf{f}) = \int\limits_{-\infty}^{\infty} a(\mathbf{x}) \frac{Z}{R} \left( 1 - i \frac{2\pi}{\lambda} R \right) \frac{\exp(i2\pi R / \lambda)}{R^2} d\mathbf{x} \,, \qquad (1.3.1)$$

where $\mathbf{x} = (x, y)$ is a vector of coordinates in the object plane, $\mathbf{f} = (f_x, f_y)$ is a vector of coordinates in the hologram plane, $Z$ is a distance between object and hologram planes, $R = \sqrt{Z^2 + (x - f_x)^2 + (y - f_y)^2}$ and $\lambda$ is the radiation wavelength. In what follows in this section we, for the sake of brevity, will use one-dimensional denotations, illustrated in Fig. 2.5.

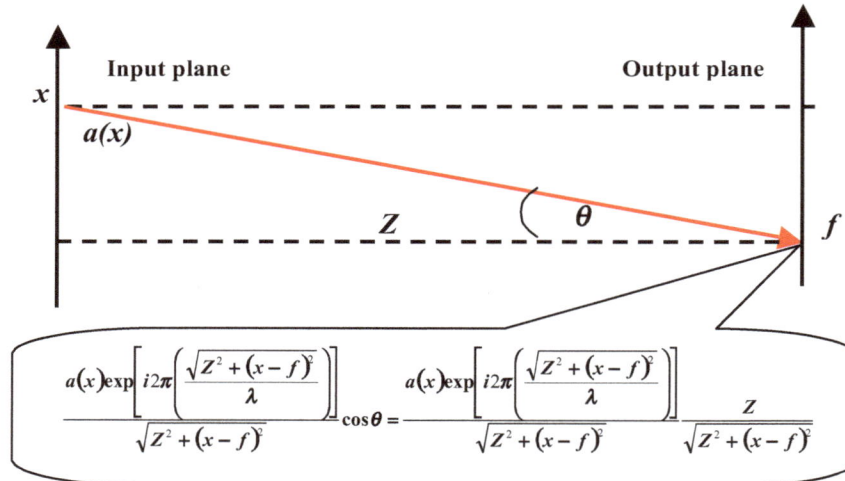

**Fig. 1.5.** Graphical illustration of the wave propagation model

Usually, an approximation to the integral (1.3.1):

$$\left( 1 - i \frac{2\pi}{\lambda} R \right) \approx i \frac{2\pi}{\lambda} R \,, \qquad (1.3.2)$$

is made and the above integral is reduced to the transform

$$\alpha(f) \approx -i\frac{2\pi}{\lambda}Z\int_{-\infty}^{\infty}a(x)\frac{\exp(i2\pi R/\lambda)}{R^2}dx \propto \int_{-\infty}^{\infty}a(x)\frac{\exp\left(i2\pi\dfrac{Z\sqrt{1+(x-f)^2/Z^2}}{\lambda}\right)}{1+(x-f)^2/Z^2}dx.$$

$$(1.3.3)$$

called ***Kirchhoff-Rayleigh-Sommerfeld integral transform (KRST )***.

Most frequently, object size and hologram size are small with respect to the object-to-hologram distance $Z$. Depending on how small they are, two other approximations to the Kirchhoff - Rayleigh-Sommerfeld integral (Eq. 1.3.3) are used, "near zone diffraction" (Fresnel) approximation:

$$\alpha(f) \propto \int_{-\infty}^{\infty}a(x)\exp\left[i\pi\frac{(x-f)^2}{\lambda Z}\right]dx \qquad (1.3.4)$$

known as ***Fresnel integral transform***, and "far zone diffraction" (Fraunhofer) approximation which is well known ***Fourier integral transform***.

$$\alpha(f) \propto \int_{-\infty}^{\infty}a(x)\exp\left(-i2\pi\frac{xf}{\lambda Z}\right)dx \qquad (1.3.5)$$

There is also a version of the Fresnel Transform called ***angular spectrum propagation*** ([4]). In this version, Fresnel Transform is treated as a convolution.

By the Fourier convolution theorem, Fresnel transform (Eq. 1.3.4) can be represented as inverse Fourier Transform

$$\alpha(f) \propto \int_{-\infty}^{\infty}\left[\int_{-\infty}^{\infty}a(x)\exp(i2\pi x\xi)dx\right]\left[\int_{-\infty}^{\infty}\exp\left(i\pi\frac{x^2}{\lambda Z}\right)\exp(i2\pi x\xi)dx\right]\exp(-i2\pi f\xi)d\xi \propto$$

$$\int_{-\infty}^{\infty}\left[\int_{-\infty}^{\infty}a(x)\exp(i2\pi x\xi)dx\right]\exp(-i\pi\lambda Z\xi^2)\exp(-i2\pi f\xi)d\xi \qquad (1.3.6)$$

of the product of signal Fourier Transform and Fourier Transform of the ***chirp function*** $\exp(i\pi x^2/\lambda Z)$, which is, in its turn, a chirp-function (Ch. 2, Appendix A2.3., Eq. A2.3.1).

## References

1.    D. Gabor, Microscopy by reconstructed waveforms, **1**, *Proc. Royal Society*, A197, 454-487, (1949)
2.    E. N. Leith, J. Upatnieks, New techniques in wavefront reconstruction, *JOSA*, 51, 1469-1473, (1961)
3.    Yu. N. Denisyuk, Photographic reconstruction of the optical properties of an object in its own scattered radiation field, *Dokl. Akad. Nauk SSSR*, **144**,1275-1279 (1962)
4.    J. Goodman, Introduction to Fourier Optics, 3-rd edition, Roberts & Company Publishers, 2005

# 2　　Holographic Transforms in Digital Computers

**Abstract:** This chapter is devoted to discrete representation of diffraction integrals. It expounds principles of the discretization and their applications to derivation of different modifications of Discrete Fourier Transforms, Discrete Fresnel Transforms and Kirchhoff-Rayleigh-Sommerfeld transforms. The chapter is supplemented with appendices that contain relevant mathematical details, description of the new algorithm for signal convolution via fast cosine transform and a summary of discrete diffraction transforms

## 2.1　　Discrete representation of transforms: principles

For numerical reconstruction of holograms recorded in different diffraction zones, as well as for synthesis of computer generated holograms, discrete representations of the diffraction integrals are needed that are suited for efficient computational implementation.

At first glance, obtaining discrete representation of integral transform is a trivial task: one has only to replace integrals by integral sums of integrand values in sampling points. However this simplistic solution that originates from tradition of numerical mathematics disregards the theory of signal discretization, and, specifically, sampling theory and due to this can't properly treat the problem of the accuracy of the discrete representation. The rigorous approach to discrete representation of continuous signal transforms is based on two principles ([1]):
- The *conformity principle* with digital representation of signals
- The *mutual correspondence principle* between continuous and discrete transformations

The conformity principle requires that digital representation of signal transformations should parallel that of signals. The mutual correspondence principle between continuous and digital transformations is said to hold if both act to transform identical input signals into identical output signals. Thus, (i) methods for representing continuous transforms digitally should account for procedures of digitizing continuous signals and reconstructing continuous signals from their digital representatives and (ii) digital processing should be specified and characterized in terms of equivalent continuous transformation that also accounts for continuous signal digitization and reconstruction as it is illustrated in Fig. 1.

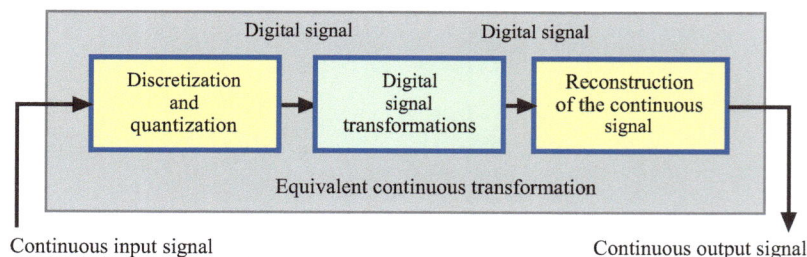

**Figure 2.1.** Mutual correspondence principle between continuous and digital signal transformations

Discretization is a process of measuring, in the discretization devices such as image scanners and hologram sensors, of coefficients of signal expansion into a series over a set of functions called *discretization basis functions*. These coefficients represent signals in computers. It is assumed that original continuous signals can be, with certain agreed accuracy, reconstructed by summation of functions called **reconstruction basis functions** with weights equal to the corresponding coefficients of signal discrete representation. The reconstruction is carried out in signal reconstruction devices such as, for instance, image displays and computer generated hologram recorders.

In digital holography and digital image processing, discretization and reconstruction basis functions that belong to a family of **"shift"**, or sampling, **basis functions** are most commonly used. All functions from this family are obtained from one "mother" function by means of its spatial shifts through multiple of a **sampling interval**. Signal discrete representation coefficients obtained for such functions are called **signal samples**.

The mathematical formulation of signal sampling and reconstruction from the sampled representation is as follows. Let $a(\mathbf{x})$ be a continuous signal as a function of spatial co-ordinates given by a vector $\mathbf{x}^{(s)}$,

$$\phi_k^{(s)}\left(\mathbf{x}^{(s)}\right)=\phi^{(s)}\left(\mathbf{x}^{(s)}-\tilde{\mathbf{k}}^{(s)}\ddot{A}\mathbf{x}^{(s)}\right) \qquad (2.1.1,\text{ a})$$

and

$$\phi_k^{(r)}\left(\mathbf{x}^{(r)}\right)=\phi^{(r)}\left(\mathbf{x}^{(r)}-\tilde{\mathbf{k}}^{(r)}\ddot{A}\mathbf{x}^{(r)}\right) \qquad (2.1.1,\text{ b})$$

be sampling and reconstruction basis functions defined in the sampling and reconstruction devices coordinates $\left\{\mathbf{x}^{(s)}\right\}$ and $\left\{\mathbf{x}^{(r)}\right\}$, respectively, $\ddot{A}\mathbf{x}^{(s)}$ and $\ddot{A}\mathbf{x}^{(r)}$ be vectors of the sampling intervals in, respectively, sampling and reconstruction devices, $\tilde{\mathbf{k}}^{(s)}=\mathbf{k}+\mathbf{u}^{(s)}$ and $\tilde{\mathbf{k}}^{(r)}=\mathbf{k}+\mathbf{u}^{(r)}$ be vectors of signal sample indices $\mathbf{k}$ biased by certain shift vectors $\mathbf{u}^{(s)}$ and $\mathbf{u}^{(r)}$. Shift vectors describe shift, in units of the corresponding sampling intervals, of the sampling grid with respect to the corresponding coordinate systems $\mathbf{x}^{(s)}$ and $\mathbf{x}^{(r)}$, such that samples with index $\mathbf{k}=0$ are assumed to have respective coordinates $\mathbf{x}^{(s)}=\mathbf{u}^{(s)}$, $\mathbf{x}^{(r)}=\mathbf{u}^{(r)}$ (Fig. 2.2).

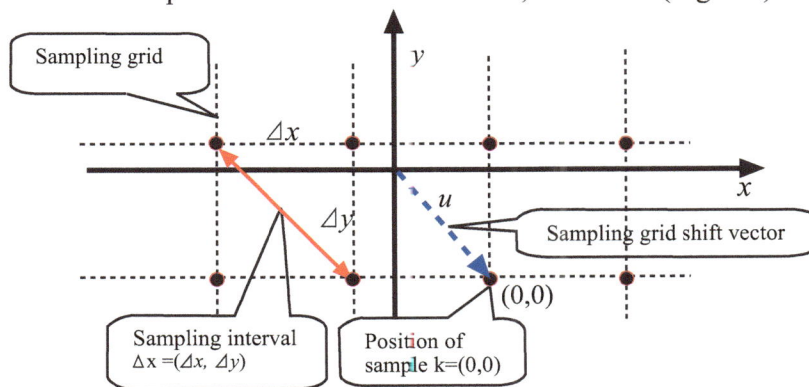

**Fig. 2.2.** Signal sampling grid in the signal coordinate system (*x,y*)

At sampling, signal samples $\{a_k\}$ are computed in sampling devices as projections of the signal onto sampling basis functions:

$$a_k = \int_X a(\mathbf{x})\phi^{(s)}\left(\mathbf{x}^{(s)} - \tilde{\mathbf{k}}^{(s)}\Delta\mathbf{x}^{(s)}\right)d\mathbf{x},$$                     (2.1.2)

assuming certain relationship between signal and sampling device coordinate systems $\{\mathbf{x}\}$ and $\{\mathbf{x}^{(s)}\}$.

Signal reconstruction from the set of their samples $\{a_k\}$ is described as signal expansion over the set of reconstruction basis functions:

$$\tilde{a}\left(\mathbf{x}^{(r)}\right) = \sum_k a_k\phi^{(r)}\left(\mathbf{x}^{(r)} - \tilde{\mathbf{k}}^{(r)}\Delta\mathbf{x}^{(r)}\right).$$                     (2.1.3)

in a reconstruction device coordinate system.

The result $\tilde{a}(\mathbf{x})$ of the signal reconstruction from its discrete representation obtained according to Eq. 2.1.3 is not, in general, identical to the initial signal $a(\mathbf{x})$. It is understood, however, that it can, in the given application, serve as a substitute for the initial signal.

Eqs. 2.1.2 and 2.1.3 model processes of hologram sampling in digital cameras and, respectively, object reconstruction in display devices. According to the conformity principle, they form the base for adequate discrete representation of signal transformations. In what follows, we describe discrete representations of optical diffraction integral transforms generated by this kind of signal representation.

## 2.2   Discrete Fourier transforms

In this section, we will provide a full derivation of discrete representations of the integral Fourier Transform. This derivation will explicitly demonstrate approximations that are done in this continuous-to-discrete conversion and will serve as an illustrative model for other diffraction transforms considered hereafter.

In order to avoid unnecessary complications we first consider a one-dimensional case. Let $\alpha(f)$ and $a(\mathbf{x})$ be, correspondingly, hologram and object wave fronts linked through the integral Fourier Transform:

$$\alpha(f) = \int_{-\infty}^{\infty} a(x)\exp\left(i2\pi\frac{xf}{\lambda Z}\right)dx$$                     (2.2.1)

In digital recording of holograms, samples of the hologram wave front are obtained as

$$\alpha_r = \int_{-\infty}^{\infty} \alpha(f)\phi^{(s)}\left(f - \tilde{r}^{(s)}\Delta f^{(s)}\right)df \tag{2.2.2}$$

where $\Delta f^{(s)}$ is the hologram sampling interval. Replacing here $\alpha(f)$ through its representation as Fourier transform of $a(\mathbf{x})$ (Eq. 2.2.1), we obtain:

$$\alpha_r = \int_{-\infty}^{\infty} \left\{ \int_{-\infty}^{\infty} a(x)\exp\left(i2\pi\frac{xf}{\lambda Z}\right)dx \right\}\phi^{(s)}\left(f - \tilde{r}^{(s)}\Delta f^{(s)}\right)df =$$

$$\int_{-\infty}^{\infty} a(x)dx \left\{ \int_{-\infty}^{\infty} \phi^{(s)}\left[f - \tilde{r}^{(s)}\Delta f^{(s)}\right]\exp\left(i2\pi\frac{xf}{\lambda Z}\right) \right\}df =$$

$$\int_{-\infty}^{\infty} a(x)\exp\left(i2\pi\frac{x\tilde{r}^{(s)}\Delta f^{(s)}}{\lambda Z}\right)dx \left\{ \int_{-\infty}^{\infty} \phi^{(s)}(f)\exp\left(i2\pi\frac{xf}{\lambda Z}\right) \right\}df =$$

$$\int_{-\infty}^{\infty} a(x)\exp\left(i2\pi\frac{x\tilde{r}^{(s)}\Delta f^{(s)}}{\lambda Z}\right)\Phi^{(s)}(x)dx , \tag{2.2.3}$$

where

$$\Phi^{(s)}(x) = \int_{-\infty}^{\infty} \phi^{(s)}(f)\exp\left(i2\pi\frac{xf}{\lambda Z}\right)df \tag{2.2.4}$$

is Fourier transform of the sampling device ***point spread function***, or its ***frequency response***, and $\left\{\tilde{r}^{(s)} = r + v^{(s)}\right\}$ are Fourier domain sample indices shifted as it was explained in Sect. 2.1. Now we can replace the object wave front $a(\mathbf{x})$ in Eq. 2.2.3 with its representation

$$a(x) = \sum_{k=0}^{N-1} a_k \phi^{(r)}\left(x - \tilde{k}^{(r)}\Delta x^{(r)}\right) \tag{2.2.5}$$

through its samples $\{a_k\}$, assuming that a certain number $N$ of its samples produced by the hologram sampling device are available, and that they are indexed from $\boldsymbol{k = 0}$ to $k = N-1$ with $\left\{\tilde{k}^{(r)} = k^{(r)} + u^{(r)}\right\}$ as correspondingly shifted indices. With this we obtain:

$$\alpha_r = \int_{-\infty}^{\infty} a(x)\exp\left(i2\pi\frac{x\tilde{r}^{(s)}\Delta f^{(s)}}{\lambda Z}\right)\Phi^{(s)}(x)dx =$$

$$\int_{-\infty}^{\infty} \left\{ \sum_{k=0}^{N-1} a_k \phi^{(r)}\left(x - \tilde{k}^{(r)}\Delta x^{(r)}\right) \right\}\exp\left(i2\pi\frac{x\tilde{r}^{(s)}\Delta f^{(s)}}{\lambda Z}\right)\Phi^{(s)}(x)dx =$$

$$\left\{ \sum_{k=0}^{N-1} a_k \exp\left( i2\pi \tilde{k}^{(r)} \tilde{r}^{(s)} \frac{\Delta x^{(r)} \Delta f^{(s)}}{\lambda Z} \right) \right\} \times$$

$$\left\{ \int_{-\infty}^{\infty} \phi^{(r)}(x) \Phi^{(s)}\left( x + \tilde{k}^{(s)} \Delta x^{(r)} \right) \exp\left( i2\pi \frac{x\tilde{r}^{(s)} \Delta f^{(s)}}{\lambda Z} \right) dx \right\}. \tag{2.2.6}$$

This expression has two multiplicands. As the discrete representation of the integral Fourier transform, only the first multiplicand that depends on signal samples is used:

$$\alpha_r = \sum_{k=0}^{N-1} a_k \exp\left( i2\pi \tilde{k}^{(r)} \tilde{r}^{(s)} \frac{\Delta x^{(r)} \Delta f^{(s)}}{\lambda Z} \right) \tag{2.2.7}$$

The second multiplicand that depends on physical parameters of sampling and reconstruction (display) devices is ignored. It is this term that, in addition to the approximative signal reconstruction from the finite number of its samples as described by Eq. 2.2.5, reflects approximative nature of the discrete representation of integral Fourier transform. The most adequate way to quantitatively evaluate implication of the approximations is to consider the point spread function and the resolving power of numerical Fourier analysis. This issue is addressed in Ch. 4.

Sampling interval in signal domain $\Delta x^{(r)}$ and the width of signal Fourier spectrum $N\Delta f^{(s)}$ defined by the selection of sampling interval in Fourier domain $\Delta f^{(s)}$ and of the number of signal samples $N$ are known to be linked by the "uncertainty relationship":

$$\Delta x^{(r)} \geq \frac{\lambda Z}{N\Delta f^{(s)}} \tag{2.2.8, a}$$

The case

$$\Delta x^{(r)} = \frac{\lambda Z}{N\Delta f^{(s)}} \tag{2.2.8, b}$$

associated with the assumption of the "band-limitedness" of signals and Nyquist-Kotelnikov-Shannon's **sampling theorem** is referred to as **cardinal sampling**. Depending on particular relationship between sampling intervals $\Delta x^{(r)}$ and $\Delta f^{(s)}$ and on sampling geometry, a number of modifications of the discrete Fourier Transforms can be obtained. They are presented in Sections 2.3.1 – 2.3.5.

### 2.2.1   *1-D direct and inverse Canonical Discrete Fourier Transforms (DFT)*

The basic and most widely accepted assumptions are that signal and its Fourier spectrum sampling intervals $\Delta x^{(r)}$ and $\Delta f^{(s)}$ satisfy the "cardinal" sampling relationship of Eq. 2.2.8, b) and that object signal and object sampling device coordinate systems as well as, those of the Fourier hologram and of the hologram sampling device, are, correspondingly, identical and object signal samples $\left\{ a_k \right\}$ as well as samples $\{\alpha_r\}$ of its Fourier hologram are positioned in such a way that samples with indices $k = 0$ and $r = 0$ are taken in signal and its Fourier hologram coordinates in points $x = 0$ and $f = 0$, respectively. In these assumptions, Eq. 2.2.7 leads to the **Canonical 1-D Discrete Fourier Transforms** (**DFT**):

$$\alpha_r = \frac{1}{\sqrt{N}} \sum_{k=0}^{N-1} a_k \exp\left( i2\pi \frac{kr}{N} \right) \qquad (2.2.9, \text{a})$$

One can show that this discrete transform is orthogonal and has inverse transform (**IDFT**):

$$a_k = \frac{1}{\sqrt{N}} \sum_{r=0}^{N-1} \alpha_r \exp\left( -i2\pi \frac{kr}{N} \right). \qquad (2.2.9, \text{b})$$

Canonical DFT plays a fundamental role in digital signal processing and, in particular, in digital holography and digital image processing thanks to the existence of Fast Fourier Transform (FFT) algorithms. With the use of FFT, computational complexity of transforms is as small as $O(\log N)$ operations per sample. All discrete transforms considered in this chapter are reducible to DFT and can be computed using FFT algorithms. Thanks to the existence of FFT algorithms, DFT is also very frequently used for fast computation of signal cyclic convolution through the following algorithm:

$$\sum_{n=0}^{N-1} a_n b_{(k-n) \bmod N} = \text{IFFT}_N \left\{ \text{FFT}_N \left( \{a_n\} \right) \bullet \text{FFT}_N \left( \{b_n\} \right) \right\}, \qquad (2.2.10)$$

where $\text{FFT}_N\{\cdot\}$ and $\text{IFFT}_N\{\cdot\}$ are operators of $N$-point FFTs applied to vectors of $N$ signal samples and symbol $\bullet$ designates component-wise multiplication of the transform results.

### 2.2.2    1D direct and inverse Shifted DFTs. Discrete Cosine Transform

Let the "cardinal" sampling relationship (2.1.8, b) between object signal and its Fourier spectrum sampling intervals $\Delta x^{(r)}$ and $\Delta f^{(s)}$ holds and object and object signal sampling device coordinate systems as well as  those of signal spectrum and the spectrum sampling device, are laterally shifted such that signal sample $\{a_0\}$ and sample $\{\alpha_0\}$ of its Fourier spectrum are taken in signal and, correspondingly, spectrum coordinates at points $x = u^{(r)}\Delta x^{(\cdot)}$ and $f = v^{(s)}\Delta f^{(s)}$, respectively as it is shown in Fig. 2.2. In this case 1-D direct and inverse **Shifted DFTs**  (**SDFT(u,v)**) are obtained from Eq. 2.2.7 ([1, 2]):

$$\alpha_r^{u,v} = \frac{1}{\sqrt{N}} \sum_{k=0}^{N-1} a_k \exp\left[ i2\pi \frac{\left(k^{(r)} + u^{(r)}\right)\left(r + v^{(s)}\right)}{N} \right],$$

$$\alpha_k^{u,v} = \frac{1}{\sqrt{N}} \sum_{r=0}^{N-1} \alpha_r^{u,v} \exp\left[-i2\pi \frac{\left(k^{(r)} + u^{(r)}\right)\left(r + v^{(s)}\right)}{N}\right] \qquad (2.2.11)$$

Superscripts (*u,v*) are assigned in Eq. 2.2.11 to signal spectral coefficients $\left\{\alpha_r^{u,v}\right\}$ and to the result of inverse transform $\left\{a_k^{u,v}\right\}$ in order to emphasize that, while signal samples $\left\{a_k\right\}$ are defined for index $k$ within the basic interval $[0, N-1]$, samples $\left\{\alpha_r^{u,v}\right\}$ and $\left\{a_k^{u,v}\right\}$ are defined outside the basic interval and their relations to the samples within the basic interval are determined by sampling sift parameters (*u,v*).

$$\alpha_{r+gN}^{u,v} = \alpha_r^{u,v} \exp\left(i2\pi g u^{(r)}\right) \; ; \; a_{k+gN}^{u,v} = a_k^{u,v} \exp\left(-i2\pi g v^{(s)}\right) , \qquad (2.2.12)$$

where $g$ is an integer number.

SDFT can obviously be reduced to DFT and, therefore, be computed using FFT algorithms. The availability in SDFT of arbitrary shift parameters enables efficient algorithms for hologram reconstruction and object signal re-sampling with sub-pixel accuracy and with the perfect discrete sinc-interpolation of sampled data ([1]). For instance, if one computes ***DFT*** of a signal $\left\{a_n\right\}$ and then takes ***IDFT***(*u*,0) of obtained spectrum, a *u*-shifted re-sampled copy $\left\{a_k^u\right\}$ of the initial signal will be obtained linked with the initial signal samples $\left\{a_n\right\}$ by the following interpolation relationship called ***discrete sinc-interpolation*** ([1]):

$$a_k^u = \sum_{n=0}^{N-1} a_n \frac{\sin\left[\pi\left(n-k+u\right)\right]}{N\sin\left[\pi\left(n-k+u\right)/N\right]} = \sum_{n=0}^{N-1} a_n \operatorname{sincd}\left[N, \pi\left(n-k+u\right)\right], \qquad (2.2.13)$$

where

$$\operatorname{sincd}\left(N;x\right) = \frac{\sin x}{N\sin x/N} \qquad (2.2.14)$$

is the discrete sinc-function (***sincd-function***).

Important special case of Shifted DFTs is ***Discrete Cosine Transform*** (DCT)

$$\alpha_r = \sum_{k=0}^{N-1} a_k \cos\left(\pi \frac{k+1/2}{N} r\right) \qquad (2.2.15)$$

One can easily show that it is ***SDFT***(1/2,0), with shift parameters 1/2 in the signal domain and 0 in the transform domain, of signals that exhibit even symmetry

($\left\{ a_k = +a_{2N-1-k} \right\}$). DCT has fast computational algorithm that belongs to the family of fast Fourier Transform algorithms ([1]).

DCT has numerous applications in image processing and digital holography. In particular, using fast DCT algorithms, one can efficiently, with the complexity of $O(\log N)$ operations per output sample, implement fast digital convolution (see Appendix A2.1). Signal even extension $\left\{ a_k = a_{2N-1-k} \right\}$ to the double length of $2N-1$ samples, which converts DFT(1/2,0) to DCT, eliminates signal potential discontinuities at borders of its initial length. This allows to substantially alleviate severe boundary effects, which are characteristic for cyclic (periodic) convolution, when convolution is implemented by processing in the domain of Discrete Fourier Transforms.

In conclusion of this section note that many other versions of SDFT with semi-integer and integer shift parameters can be easily derived for different types of signal and its spectrum symmetries.

### 2.2.3   1-D Scaled DFT

If one assumes that sampling rate of either signal samples or spectrum samples or both is $\sigma$-times the "cardinal" sampling rate ($\Delta x^{(r)} = \lambda Z / \sigma N \Delta f^{(s)}$), and that signal and its Fourier hologram samples $\left\{ a_0 \right\}$ and $\left\{ \alpha_0 \right\}$ are positioned with shifts $\left( u^{(r)}, v^{(s)} \right)$ with respect to the origins of the corresponding signal and spectrum coordinate systems, *Scaled DFT (ScDFT)*

$$\alpha_r^\sigma = \frac{1}{\sqrt{\sigma N}} \sum_{k=0}^{N-1} a_k \exp\left( i 2\pi \frac{\tilde{k}\tilde{r}}{\sigma N} \right) , \left\{ \tilde{r}^{(s)} = r + v^{(s)} \right\} ; \left\{ \tilde{k}^{(r)} = k^{(r)} + u^{(r)} \right\} \quad (2.2.16, \text{a})$$

is obtained. Modification of this transform for zero shift parameters is also known under the names of *"chirp z-transform"* ([3,4]) and *"fractional discrete Fourier transform"*. The first name is associated with the way to compute it efficiently (see below Eq. 2.2.18). The second name assumes that it is a discretized version of the fractional integral Fourier transform ([5,6]]) that has found some applications in optics and quantum mechanics. We prefer the name "Scaled DFT" because it is more intuitive, refers to its physical interpretation and leads to a unified nomenclature of discrete Fourier transforms (shifted, scaled, rotated, scaled and rotated, affine) introduced in this chapter.

*Scaled DFT* has its inverse only if $\sigma N$ is an integer number ($\sigma N \in \text{Æ}$) and $\sigma > 1$ (see Appendix A2.2.). In this case, the inverse ScDFT is defined as

$$a_k^\sigma = \frac{1}{\sqrt{\sigma N}} \sum_{r=0}^{\sigma N-1} \alpha_r \left( -i 2\pi \frac{\tilde{k}\tilde{r}}{\sigma N} \right) = \begin{cases} a_k, & k = 0,1,...N-1 \\ 0, & k = N, N+1, ..., \sigma N - 1 \end{cases} \quad (2.2.16,\text{b})$$

For computational purposes, it is convenient to represent ScDFT as a cyclic convolution

$$\alpha_r^\sigma = \frac{1}{\sqrt{\sigma N}}\sum_{k=0}^{N-1} a_k \exp\left(i2\pi\frac{\tilde{k}\tilde{r}}{\sigma N}\right)=$$

$$\frac{\exp\left(i\pi\frac{\tilde{r}^2}{\sigma N}\right)}{\sqrt{\sigma N}}\sum_{k=0}^{N-1}\left[a_k \exp\left(i\pi\frac{\tilde{k}^2}{\sigma N}\right)\right]\exp\left[-i\pi\frac{(\tilde{k}-\tilde{r})^2}{\sigma N}\right],\qquad(2.2.17)$$

that can be computed using FFT algorithm as following:

$$\alpha_r^{(\sigma)} = \text{IFFT}_{\lfloor\sigma N\rfloor}\left\{ZP_{\lfloor\sigma N\rfloor}\left[\text{FFT}_N\left\{a_k \exp\left(i\pi\frac{\tilde{k}^2}{\sigma N}\right)\right\}\right]\bullet\text{FFT}_{\lfloor\sigma N\rfloor}\left\{\exp\left(-i\pi\frac{\tilde{n}^2}{\sigma N}\right)\right\}\right\},$$

$$(2.2.18)$$

where $\text{FFT}_M\{\cdot\}$ and $\text{IFFT}_M\{\cdot\}$ denote $M$-point direct and inverse FFTs, $\lfloor\sigma N\rfloor$ is an integer number defined by the inequality $\sigma N \le \lfloor\sigma N\rfloor < \sigma N+1$, and $ZP_M[\cdot]$ is a zero-padding operator. If $\sigma>1$, operator $ZP_{\lfloor\sigma N\rfloor}[.]$ pads the array of $N$ samples with zeros to the array of $\lfloor\sigma N\rfloor$ samples. If $\sigma<1$, it cuts array of $N$ samples to size of $\lfloor\sigma N\rfloor$ samples.

The availability in *ScDFT*s with the arbitrary scale parameter enables signal discrete-sinc-interpolated re-sampling and Fourier hologram reconstruction in an arbitrary scale. For instance, if one computes canonical DFT of a signal and then applies to the obtained spectrum *ScIDFT* with a scale parameter $\sigma$, discrete sinc-interpolated samples of the initial signal in a scaled coordinate system will be obtained:

$$a_k^\sigma = \sum_{n=0}^{N-1} a_n \text{sincd}\left[N;\pi\left(n-k/\sigma\right)\right],\qquad(2.2.19)$$

If $\sigma>1$, signal $\{a_k\}$ is sub-sampled (its discrete representation is zoomed in) with discrete sinc-interpolation and the sub-sampled signal retains the initial signal bandwidth. If $\sigma<1$, signal $\{a_k\}$ is down-sampled (decimated). For signal down-sampling, appropriate signal low-pass filtering required to avoid aliasing artifacts is automatically carried out by the imbedded zero-padding operator $ZP_{\lfloor\sigma N\rfloor}$.

### 2.2.4   *2-D Canonical Separable DFTs.*

In order to introduce discrete representation of 2-D integral transforms, one should specify a 2-D sampling grid. Most frequently, rectangular sampling grid (Fig. 2.2) is assumed for sampled representation of 2-D signals. In this case, for 2-D integral Fourier Transform, the following separable 2-D canonical direct and inverse DFTs:

$$\alpha_{r,s} = \frac{1}{\sqrt{N_1 N_2}}\sum_{k=0}^{N_1-1}\exp\left(i2\pi\frac{kr}{N_1}\right)\sum_{l=0}^{N_2-1}a_{k,l}\exp\left(i2\pi\frac{ls}{N_2}\right);\qquad(2.2.20,a)$$

$$a_{k,l} = \frac{1}{\sqrt{N_1 N_2}} \sum_{r=0}^{N_1-1} \exp\left(-i2\pi \frac{kr}{N_1}\right) \sum_{s=0}^{N_2-1} \alpha_{r,s} \exp\left(-i2\pi \frac{ls}{N_2}\right). \qquad (2.2.20, b)$$

are obtained under the assumption that object signal and its hologram sampling and reconstruction are performed in rectangular sampling grids (row-wise, column-wise) collinear with the object coordinate system. Here $N_1$ and $N_2$ are dimensions of 2-D arrays of signal and its Fourier spectrum samples. In separable 2-D DFTs, 1-D Shifted and Scaled DFTs can also be used to enable signal 2-D separable discrete sinc-interpolated resampling and re-scaling.

### 2.2.5   2-D Rotated and Affine DFTs.

In 2-D case, a natural generalization of 1-D Shifted and Scaled DFTs is 2-D **Affine DFT** (**AffDFT**). **AffDFT** is obtained in the assumption that either signal or its spectrum sampling are carried out in affine transformed, with respect to signal/spectrum coordinate systems $(x, y)$, coordinates $(\tilde{x}, \tilde{y})$:

$$\begin{bmatrix} x \\ y \end{bmatrix} = \begin{bmatrix} A & B \\ C & D \end{bmatrix} \begin{bmatrix} \tilde{x} \\ \tilde{y} \end{bmatrix} \qquad (2.2.21)$$

With $\qquad \sigma_A = \lambda Z / N_1 A \Delta\tilde{x} \Delta f_x$; $\qquad \sigma_B = \lambda Z / N_2 B \Delta\tilde{y} \Delta f_x$; $\qquad \sigma_C = \lambda Z / N_1 C \Delta\tilde{x} \Delta f_y$,

$\sigma_D = \lambda Z / N_2 D \Delta\tilde{y} \Delta f_y$, where $\Delta\tilde{x}$, $\Delta\tilde{y}$, $\Delta f_x$ and $\Delta f_y$ are object and its Fourier hologram sampling intervals in object $(\tilde{x}, \tilde{y})$ and hologram $(f_x, f_y)$ planes, **AffDTF** is defined as

$$\alpha_{r,s} = \sum_{k=0}^{N_1-1} \sum_{l=0}^{N_2-1} a_{k,l} \exp\left[ i2\pi \left( \frac{\tilde{r}\tilde{k}}{\sigma_A N_1} + \frac{\tilde{s}\tilde{k}}{\sigma_C N_1} + \frac{\tilde{r}\tilde{l}}{\sigma_B N_2} + \frac{\tilde{s}\tilde{l}}{\sigma_D N_2} \right) \right], \qquad (2.2.22)$$

where $\{\tilde{k}, \tilde{l}\}$ and $\{\tilde{r}, \tilde{s}\}$ are biased (shifted) sampling indices in object and hologram planes.

A special case of affine transforms is rotation. For the rotation angle $\theta$ rotated with respect to each other coordinate systems are connected by the relationship:

$$\begin{bmatrix} x \\ y \end{bmatrix} = \begin{bmatrix} \cos\theta & \sin\theta \\ -\sin\theta & \cos\theta \end{bmatrix} \begin{bmatrix} \tilde{x} \\ \tilde{y} \end{bmatrix} \qquad (2.2.23)$$

With $N_1 = N_2 = N$, $\Delta\tilde{x} = \Delta\tilde{y} = \Delta x$, $\Delta f_x = \Delta f_y = \Delta f$, and $\Delta x \Delta f = \lambda Z / N$ (cardinal sampling), 2-D **Rotated DFT** (**RotDFT**) are obtained as

$$\alpha_{r,s}^{\theta} = \frac{1}{\sigma N} \sum_{k=0}^{N-1} \sum_{l=0}^{N-1} a_{k,l} \exp\left[ i2\pi \left( \frac{\tilde{k}\cos\theta + \tilde{l}\sin\theta}{N} \tilde{r} - \frac{\tilde{k}\sin\theta - \tilde{l}\cos\theta}{N} \tilde{s} \right) \right]. \quad (2.2.24)$$

An obvious generalization of RotDFT is **Rotated and Scaled DFT** (**RotScDFT**):

$$\alpha_{r,s}^{\theta} = \frac{1}{\sigma N}\sum_{k=0}^{N-1}\sum_{l=0}^{N-1} a_{k,l}\exp\left[i2\pi\left(\frac{\tilde{k}\cos\theta+\tilde{l}\sin\theta}{\sigma N}\tilde{r}-\frac{\tilde{k}\sin\theta-\tilde{l}\cos\theta}{\sigma N}\tilde{s}\right)\right] \quad (2.2.25)$$

that assumes 2-D signal sampling in $\theta$-rotated and $\sigma$-scaled coordinate systems. Similarly to **ScDFT**, **RotScDFT** can be reduced to 2-D cyclic convolution using identities:

$$\alpha_{r,s}^{\theta} = \frac{1}{\sigma N}\sum_{k=0}^{N-1}\sum_{l=0}^{N-1} \tilde{a}_{k,l}\exp\left[i2\pi\left(\frac{\tilde{k}\cos\theta+\tilde{l}\sin\theta}{\sigma N}\tilde{r}-\frac{\tilde{k}\sin\theta-\tilde{l}\cos\theta}{\sigma N}\tilde{s}\right)\right] =$$
$$\frac{1}{\sigma N}\sum_{k=0}^{N-1}\sum_{l=0}^{N-1} \tilde{a}_{k,l}\exp\left[i2\pi\left(\frac{\tilde{k}\tilde{r}+\tilde{l}\tilde{s}}{\sigma N}\cos\theta+\frac{\tilde{l}\tilde{r}-\tilde{k}\tilde{s}}{\sigma N}\sin\theta\right)\right] \quad (2.2.26)$$

$$2\left(\tilde{k}\tilde{r}+\tilde{l}\tilde{s}\right) = \tilde{r}^2+\tilde{k}^2-\tilde{s}^2-\tilde{l}^2-\left(\tilde{r}-\tilde{k}\right)^2-\left(\tilde{s}+\tilde{l}\right)^2 ; \quad (2.2.27)$$

$$2\left(\tilde{l}\tilde{r}-\tilde{k}\tilde{s}\right) = 2\tilde{k}\tilde{l}-2\tilde{r}\tilde{s}+2\left(\tilde{r}-\tilde{k}\right)\left(\tilde{s}+\tilde{l}\right). \quad (2.2.28)$$

Inserting Eqs. 2.2.27 and 2.2.28 into Eq. 2.2.25, obtain:

$$\alpha_{r,s}^{\theta} = \frac{1}{\sigma N}\sum_{r=0}^{N-1}\sum_{s=0}^{N-1} \tilde{a}_{k,l}\exp\left[i2\pi\left(\frac{\tilde{k}\tilde{r}+\tilde{l}\tilde{s}}{\sigma N}\cos\theta+\frac{\tilde{l}\tilde{r}-\tilde{k}\tilde{s}}{\sigma N}\sin\theta\right)\right] =$$
$$\frac{1}{\sigma N_a}\sum_{k=0}^{N-1}\sum_{l=0}^{N-1} \tilde{a}_{k,l}\exp\left[i\pi\frac{\tilde{r}^2+\tilde{k}^2-\tilde{s}^2-\tilde{l}^2-\left(\tilde{r}-\tilde{k}\right)^2-\left(\tilde{s}+\tilde{l}\right)^2}{\sigma N}\cos\theta\right]\times$$
$$\exp\left[-i2\pi\frac{2\tilde{k}\tilde{l}-2\tilde{r}\tilde{s}+2\left(\tilde{r}-\tilde{k}\right)\left(\tilde{s}+\tilde{l}\right)}{\sigma N}\sin\theta\right] =$$
$$\left\{\frac{1}{\sigma N}\sum_{k=0}^{N-1}\sum_{l=0}^{N-1}\left(\alpha_{r,s}A_{r,s}\right)\mathrm{ChF}\left(\tilde{s}+\tilde{l},\tilde{r}-\tilde{k}\right)\right\}\exp\left[-i\pi\frac{\left(\tilde{r}^2-\tilde{s}^2\right)\cos\theta-2\tilde{r}\tilde{s}\sin\theta}{\sigma N}\right],$$

$$(2.2.29)$$

where

$$\mathrm{ChF}\left(\tilde{s}+\tilde{l},\tilde{r}-\tilde{k}\right) = \exp\left[i\pi\frac{\left(\tilde{s}+\tilde{l}\right)^2\cos\theta-\left(\tilde{r}-\tilde{k}\right)^2\cos\theta-2\left(\tilde{r}-\tilde{k}\right)\left(\tilde{s}+\tilde{l}\right)\sin\theta}{\sigma N}\right];$$

$$(2.2.30, a)$$

and

$$A_{r,s} = \left\{\exp\left[-i\pi\frac{\left(\tilde{r}^2-\tilde{s}\right)^2\cos\theta+2\tilde{r}\tilde{s}\sin\theta}{\sigma N}\right]\right\}. \quad (2.2.30, b)$$

The convolution defined by Eq. 2.2.29 can be efficiently computed using FFT as:

$$\left\{\tilde{a}_{k,l}\right\} = \mathrm{IFFT2}_{\lfloor\sigma N\rfloor}\left\{\mathrm{FFT2}_{\lfloor\sigma N\rfloor}\left[\mathrm{ZP}_{\lfloor N\sigma\rfloor}\left[\mathrm{FFT2}_{N}\left(a_{k,l}\right)\right]\bullet \mathrm{A}_{r,s}\right]\bullet \mathrm{FFT2}_{\lfloor\sigma N\rfloor}\left[\mathrm{ChF}\left(r,s\right)\right]\right\},$$

$$(2.2.31)$$

where $\mathrm{FFT2}_{\lfloor N\sigma\rfloor}[\cdot]$ and $\mathrm{IFFT2}_{\lfloor N\sigma\rfloor}[\cdot]$ are operators of direct and inverse $\lfloor N\sigma\rfloor$-point 2D FFT , $\lfloor N\sigma\rfloor$ is the smallest integer larger then $N\sigma$ , $\mathrm{ZP}_{\lfloor N\sigma\rfloor}[\cdot]$ is a 2-D zero-padding operator. For $\sigma > 1$, it pads array of $N \times N$ samples with zeros to the array of $\lfloor\sigma N\rfloor \times \lfloor\sigma N\rfloor$ samples with $\lfloor\sigma N\rfloor$ defined, as above, by the inequality $\sigma N \le \lfloor\sigma N\rfloor < \sigma N + 1$. For $\sigma < 1$, the zero-padding operator cuts array of $N \times N$ samples to the size of $\lfloor\sigma N\rfloor \times \lfloor\sigma N\rfloor$ samples.

## 2.3    Discrete Fresnel transforms

In this section, we outline discrete transforms and their fast algorithms that can be used for numerical evaluation of the "near zone" (Fresnel) diffraction integral. Fresnel integral transform for signal $a(x)$ was defined in Sect. 2.2 as:

$$\alpha(f) = \int_{-\infty}^{\infty} a(x)\exp\left[i\pi\frac{(x-f)^2}{\lambda Z}\right]dx \qquad (2.3.1)$$

Similarly to the case of Fourier integral transform, for discrete representation of Fresnel integral transform the following expression is used:

$$\alpha_r = \sum_{k=0}^{N-1} a_k \exp\left[i2\pi\frac{\left(\tilde{k}^{(r)}\Delta x^{(r)} - \tilde{r}^{(s)}\Delta f^{(s)}\right)^2}{\lambda Z}\right], \qquad (2.3.2)$$

where, as above, $\left\{\tilde{k}^{(r)}\right\}$ , $\left\{\tilde{r}^{(s)}\right\}$ and $\left(\Delta x^{(r)}, \Delta f^{(s)}\right)$, are "shifted" sample indices and sampling intervals in signal and transform domains, correspondingly. For further derivations, it is convenient to rewrite Eq. 2.3.2 as:

$$\alpha_r = \sum_{k=0}^{N-1} a_k \exp\left[i2\pi\frac{\left(\tilde{k}^{(r)}\Delta x^{(r)}\big/\Delta f^{(s)} - \tilde{r}^{(s)}\Delta f^{(s)}\big/\Delta x^{(r)}\right)^2\left(\Delta f^{(s)}\right)^2}{\lambda Z}\frac{\Delta x^{(r)}}{\Delta f^{(s)}}\right] =$$

$$\sum_{k=0}^{N-1} a_k \exp\left[i2\pi\frac{\left(\tilde{k}^{(r)}\Delta x^{(r)}\big/\Delta f^{(s)} - \tilde{r}^{(s)}\Delta f^{(s)}\big/\Delta x^{(r)}\right)^2\left(\Delta f^{(s)}\right)^2}{N}\frac{N\Delta x^{(r)}}{\Delta f^{(s)}\lambda Z}\right], (2.3.3)$$

Depending on specific relationships between sampling intervals in signal and Fresnel transform domain and on sampling shift parameters, different modifications of discrete representations of Fresnel integral transform can be obtained. They are described in Sects. 2.3.1 through 2.3.7 that follow.

### 2.3.1 Canonical Discrete Fresnel Transform

Similarly to Canonical DFT, direct and inverse *Canonical Discrete Fresnel Transforms* (*CDFrT*)

$$\alpha_r = \frac{1}{\sqrt{N}} \sum_{k=0}^{N-1} a_k \exp\left[ i\pi \frac{(k\mu - r/\mu)^2}{N} \right] \qquad (2.3.4,\ a)$$

and

$$a_k = \frac{1}{\sqrt{N}} \sum_{r=0}^{N-1} \alpha_r \exp\left[ -i\pi \frac{(k\mu - r/\mu)^2}{N} \right] \qquad (2.3.4,\ b)$$

are obtained as a discrete representation of the integral Fresnel transform in the assumption of the cardinal sampling relationship $\Delta x^{(r)} = \lambda Z / N \Delta f^{(s)}$ between sampling intervals in signal and transform domains and of zero shifts of object and hologram sampling grids with respect to the corresponding coordinate systems. Parameter $\mu^2$ in this formula defined as

$$\mu^2 = \frac{\Delta x^{(r)}}{\Delta f^{(s)}} = \frac{\lambda Z}{N \Delta f^2} \qquad (2.3.5)$$

plays in *DFrT* a role of a distance (focusing) parameter.

CDFrT can be readily expressed via DFT:

$$\alpha_r = \frac{1}{\sqrt{N}} \left\{ \sum_{k=0}^{N-1} \left[ a_k \exp\left( i\pi \frac{k^2 \mu^2}{N} \right) \right] \exp\left( -i2\pi \frac{kr}{N} \right) \right\} \exp\left( i\pi \frac{r^2}{\mu^2 N} \right) \qquad (2.3.6)$$

of signal and its *DFrT* spectrum multiplied by "chirp"-functions $\exp\left( i\pi \frac{k^2 \mu^2}{N} \right)$ and

$\exp\left( i\pi \frac{r^2}{\mu^2 N} \right)$ , correspondingly. Therefore, *DFrT* can be computed using FFT algorithms. In numerical reconstruction of Fresnel holograms, this method for computing DFrT is known as the "*Fourier reconstruction algorithm*".

### 2.3.2 Shifted Discrete Fresnel Transforms

If one assumes cardinal sampling condition and arbitrary shift parameters in object and/or its hologram sampling, *Shifted Discrete Fresnel Transforms* (*SDFrT*)

$$\alpha_r^{(\mu,w)} = \frac{1}{\sqrt{N}} \sum_{k=0}^{N-1} a_k \exp\left[ -i\pi \frac{\left(k\mu - r/\mu + w\right)^2}{N} \right]$$

(2.3.7)

are obtained with parameter $w$ as a joint shift parameter that unifies shifts $u^{(r)}$ and $v^{(s)}$ of sampling grids in object and hologram planes:

$$w = u^{(r)}/\mu - v^{(s)}\mu .$$

(2.3.8)

### 2.3.3 Focal Plane Invariant Discrete Fresnel Transform

Because shift parameter $w$ in Shifted DFrT is a combination of shifts in object and hologram planes, shift in object plane causes a corresponding shift in Fresnel hologram plane, which, however, depends, according to Eq. 2.5.8, on the focusing parameter $\mu$. One can break this interdependence if, in the definition of the discrete representation of integral Fresnel transform, impose a symmetry condition

$\alpha_r^{(\mu,w)} = \alpha_{N-r}^{(\mu,w)}$ for the transform $\alpha_r^{(\mu,w)} = \exp\left[ -i\pi \frac{\left(r/\mu - w\right)^2}{N} \right]$ of a point source

$\delta(k) = 0^k$, $k = 0,1,...,N-1$. This condition is satisfied when $w = N/2\mu$ , and SDFrT for such a shift parameter takes form:

$$\alpha_r^{\left(\mu,\frac{N}{2\mu}\right)} = \frac{1}{\sqrt{N}} \sum_{k=0}^{N-1} a_k \exp\left\{ -i\pi \frac{\left[k\mu - (r - N/2)/\mu\right]^2}{N} \right\} .$$

(2.3.9)

We refer to this discrete transform as ***Focal Plane Invariant Discrete Fresnel Transform (FPIDFrT)***.

In numerical reconstruction of holograms, position of the reconstructed object in the output sampling grid depends on the object-hologram distance when canonical ***DFrT*** is used. ***FPIDFrT*** defined by Eq. (2.3.9) allows keeping position of reconstructed objects invariant to the object-hologram distance ([7]), which might be useful in applications. In particular, invariance of the reconstruction object position with respect to the distance parameter can ease automatic object focusing and usage of pruned FFT algorithms ([1]) in reconstruction of a part of the field of view.

Figs. 2.3 illustrates results of reconstruction of a digital hologram in different focal planes using Canonical and Focal Plane Invariant Discrete Fresnel Transforms.

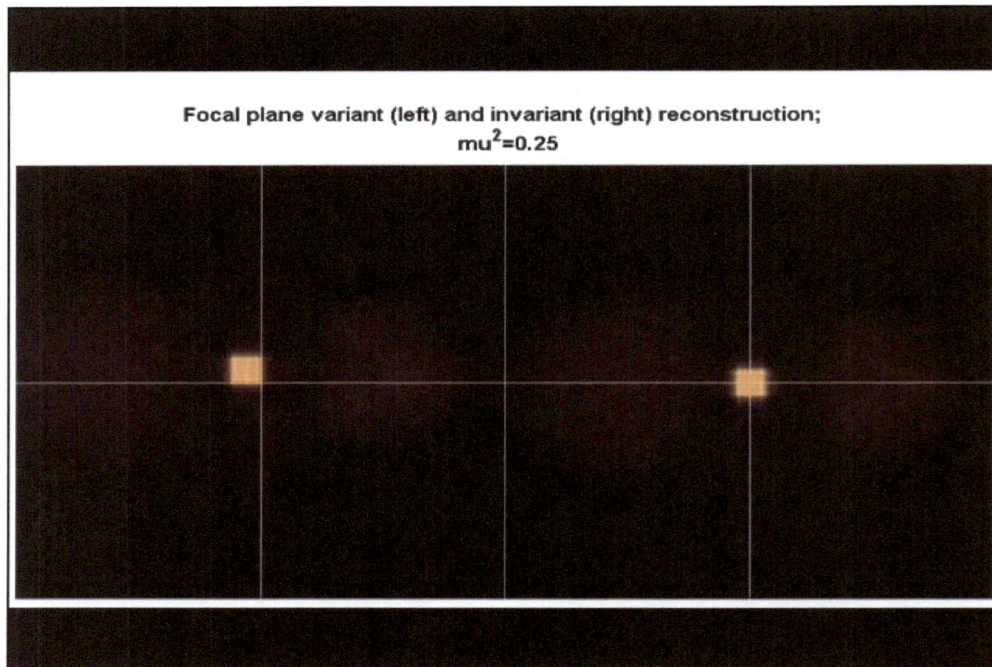

**Fig. 2.3.** Digitally recorded hologram (top) and its reconstruction on different depth using Canonical Discrete Fresnel Transform (bottom, left) and Focal Plane Invariant Discrete Fresnel Transform (bottom, right).

### 2.3.4 Partial Discrete Shifted Fresnel Transform

When, in hologram reconstruction, only intensity of the object wavefront is required, direct and inverse *Partial Discrete Fresnel Transforms* (PDFrT)

$$\widehat{\alpha}_r^{(\mu,w)} = \frac{1}{\sqrt{N}} \sum_{k=0}^{N-1} a_k \exp\left(-i\pi \frac{k^2\mu^2}{N}\right) \exp\left[i2\pi \frac{k(r-w\mu)}{N}\right]; \qquad (2.3.10,\text{ a})$$

$$a_k^{(\mu,w)} = \frac{1}{\sqrt{N}} \sum_{r=0}^{N-1} \widehat{\alpha}_r^{(\mu,w)} \exp\left[-i2\pi \frac{(r-w\mu)k}{N}\right] \exp\left(i\pi \frac{k^2\mu^2}{N}\right). \qquad (2.3.10,\text{ b})$$

can be used as discrete representations of the integral Fresnel Transform. They are obtained by removing, from Eqs. 2.3.7 exponential phase terms that do not depend on signal sampling index $k$. As one can see from Eqs. 2.3.10, direct and inverse PDFrT are essentially versions of SDFT.

### 2.3.5 Invertibility of Discrete Fresnel Transforms and frincd-function

For shifted shift and focusing parameters, Discrete Fresnel Transforms are invertible orthogonal transforms. If, however, one computes, for a discrete signal $\{a_k\}$, $k = 0,1,...,N-1$, direct SDFrT with parameters $(\mu_+, w_+)$ and then inverts it with inverse SDFrT with parameters $(\mu_-, w_-)$, the following result, different from $\{a_k\}$, will be obtained:

$$a_k^{(\mu^\pm, w^\pm)} = \frac{\exp\left[-i\pi\frac{(k\mu_- + w_-)^2}{N}\right]}{N} \times$$

$$\sum_{n=0}^{N-1} a_n \exp\left[i\pi\frac{(n\mu_+ + w_+)^2}{N}\right] \mathrm{frincd}\left(N; q; n - k + \overline{w}_\pm + qN/2\right) \qquad (2.3.11)$$

where

$$\mathrm{frincd}(N; q; x) = \frac{1}{N}\sum_{r=0}^{N-1} \exp\left(i\pi\frac{qr^2}{N}\right)\exp\left(-i2\pi\frac{xr}{N}\right). \qquad (2.3.12, a)$$

and $q = 1/\mu_+^2 - 1/\mu_-^2$; $\overline{w}_\pm = \overline{w}_+/\mu_+ - \overline{w}_-/\mu_-$. It is a function analogous to the sincd-function of the DFT (Eq. 2.1.14) and identical to it when $q = 0$. We refer to this function as to **frincd-function**. In numerical reconstruction of holograms, frincd-function is the convolution kernel that links object and its "out of focus" reconstruction.

A focal plane invariant version of frincd- function is

$$\overline{\mathrm{frincd}}(N; q; x) = \frac{1}{N}\sum_{r=0}^{N-1} \exp\left[i\pi\frac{qr(r - N)}{N}\right]\exp\left(-i2\pi\frac{xr}{N}\right) \qquad (2.3.12, b)$$

Fig. 2.4 illustrates behavior of absolute values of function $\overline{\mathrm{frincd}}(N; q; x)$ for different values of $q$ in the range $0 \le q \le 1$. In Fig. 2.5, absolute values of function $\sqrt{q}\,\overline{\mathrm{frincd}}(N; q; x)$ for $q$ in the range $0 \le q \le 2.5$ are shown as an image in coordinates $(q,x)$ to demonstrate aliasing artifacts that appear $q > 1$.

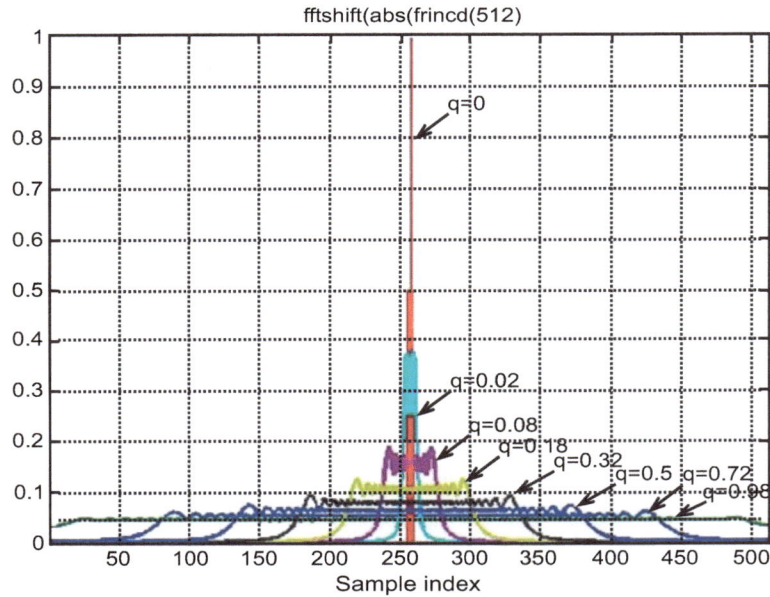

**Fig.2.4.** Absolute values of function $\overline{\text{frincd}}(N;q;x)$ for several values of $q$.

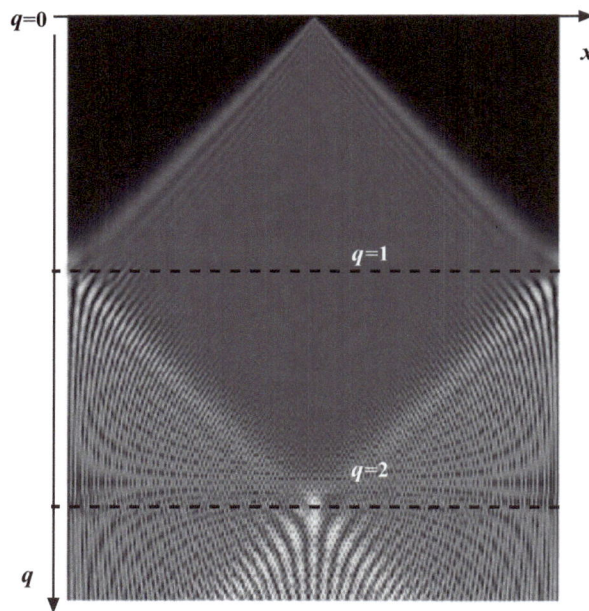

**Fig. 2.5.** Function $\sqrt{q}\,\overline{\text{frincd}}\left(N;q;x\right)$ represented, in coordinates ($q,x$), as an image with lightness proportional to its absolute values. Note aliasing artifacts for $q>1$ that culminate when $q \geq 2$

As one can see from Eqs. 2.3.12, frincd-function is DFT of a chirp-function. It is known that integral Fourier transform of a chirp-function is also a chirp-function. In Appendix A2.1 it is shown that, in distinction from the continuous case, frincd-function can only be approximated by a chirp-function:

$$\text{frincd}\left(N;\pm q;x\right) \cong \sqrt{\frac{\pm i}{Nq}}\exp\left[\mp i\pi\frac{x^2}{qN}\right]\text{rect}\left[\frac{x}{q(N-1)}\right] \tag{2.3.13, a}$$

For $q=1$ and integer $x$, it is reduced to an exact chirp-function

$$\text{frincd}\left(N;1;x\right) = \sqrt{\frac{i}{N}}\exp\left(-i\pi\frac{x^2}{N}\right). \qquad\qquad x\in\mathbb{Z} \tag{2.3.13, b}$$

As it was already indicated, for $q=0$ frincd-function reduces to sincd-function:

$$\text{frincd}\left(N;0;x\right) = \text{sincd}\left(N;\pi x\right)\exp\left(-i\pi\frac{N-1}{N}x\right) \tag{2.3.13, c}$$

### 2.3.6 Convolutional Discrete Fresnel Transform

For small distances $Z$, canonical Discrete Fresnel Transform and its above-described versions face aliasing problems when focusing parameter $\mu^2 = \lambda Z/N\Delta f^2$ is becoming less then 1, or correspondingly, parameter **$q$** of the frincd-function is larger then 1. For such cases, discrete representation of integral Fresnel transform can be used that is built on the base of the angular spectrum propagation version of the integral Fresnel Transform (Eq. 1.3.6):

$$\alpha(f) \propto \int\limits_{-\infty}^{\infty}\left[\int\limits_{-\infty}^{\infty}a(x)\exp\left(i2\pi x\xi\right)dx\right]\exp\left(-i\pi\lambda Z\xi^2\right)\exp\left(-i2\pi f\xi\right)d\xi =$$

$$\int\limits_{-\infty}^{\infty}\int\limits_{-\infty}^{\infty}a(x)\exp\left[i2\pi(x-f)\xi\right]\exp\left(-i\pi\lambda Z\xi^2\right)dx\,d\xi \tag{2.3.14}$$

and in the deliberate assumption that sampling intervals $\Delta x^{(r)}$ and $\Delta f^{(s)}$ of the object signal and its Fresnel transform are identical: $\Delta x^{(r)} = \Delta f^{(s)}$. In this assumption, the following version of discrete Fresnel Transform referred to as **Convolutional Discrete Fresnel Transform** (**ConvDFrT**) is obtained for object and hologram sampling shift parameters $u^{(r)}$ and $v^{(s)}$ :

$$\alpha_r = \frac{1}{N}\sum_{s=0}^{N-1}\left[\sum_{k=0}^{N-1}a_k\exp\left(i2\pi\frac{k-r+w}{N}s\right)\right]\exp\left(-i\pi\frac{\mu^2 s^2}{N}\right) =$$

$$\frac{1}{N}\sum_{k=0}^{N-1}a_k\left[\sum_{s=0}^{N-1}\exp\left(-i\pi\frac{\mu^2 s^2}{N}\right)\exp\left(i2\pi\frac{k-r+w}{N}s\right)\right], \tag{2.3.15, a}$$

or

$$\alpha_r = \sum_{k=0}^{N-1}a_k\,\text{frincd}^*\left(N;\mu^2;k-r+w\right) \tag{2.3.15, b}$$

where $w = u^{(r)} - v^{(s)}$ and asterisk denotes complex conjugate

ConvDFrT, similarly to DFTs and DFrTs, is an orthogonal transform with inverse ***ConvDFrT*** defined as

$$a_k = \frac{1}{N}\sum_{s=0}^{N-1}\left[\sum_{r=0}^{N-1}\alpha_r \exp\left(-i2\pi\frac{k-r+w}{N}s\right)\right]\exp\left(i\pi\frac{\mu^2 s^2}{N}\right) =$$

$$\sum_{r=0}^{N-1}\alpha_r \, \mathrm{frincd}\left(N;\mu^2;k-r+w\right). \tag{2.3.15, c}$$

When $\mu^2 = 0$ and $w = 0$, ***ConvDFrT*** degenerates to the identical transform, which is only natural because in this case object plane and hologram planes coincide.

Although ConvDFrT as a discrete transform can be inverted for any $\mu^2$, in numerical reconstruction of physical holograms it can be recommended only for $\mu^2 = \lambda Z/N\Delta f^2 \leq 1$. As it will be shown in Ch. 4, if $\mu^2 = \lambda Z/N\Delta f^2 > 1$, aliasing appears in form of overlapping periodical copies of the reconstruction result.

### 2.3.7 2-D Discrete Fresnel Transforms. Scaled and rotated transforms

2-D discrete Fresnel Transforms are usually defined as separable row-column transforms. For instance, canonical 2-D Fresnel Transform for 2-D signals $\{a_{k,l}\}$ of $N_1 \times N_2$ samples is defined as

$$\alpha_{r,s}^{(\mu,0,0)} = \frac{1}{\sqrt{N_1 N_2}}\sum_{k=0}^{N_1-1}\exp\left[-i\pi\frac{(k\mu - r/\mu)^2}{N_1}\right]\sum_{l=0}^{N_2-1}a_{k,l}\exp\left[-i\pi\frac{(l\mu - s/\mu)^2}{N_2}\right]. \tag{2.3.16}$$

As the last stage in the algorithmic implementations of all versions of discrete Fresnel transforms is Discrete Fourier Transform implemented via FFT algorithm, scaled and rotated modification of DFT can be applied at this stage. This will enable scaling and/or rotation of the transform result, if they are required when using Discrete Fresnel transforms for numerical reconstruction of digitally recorded holograms. For instance, scaled reconstruction is required in reconstruction of hologram of the same object recorded in different wavelength, such as color holograms.

## 2.4    Discrete Kirchhoff-Rayleigh-Sommerfeld transforms

Using the above-described transform discretization principles and assuming, as in the case of the Convolutional Fresnel Transform, identical sampling intervals $\Delta x^{(r)}$ and $\Delta f^{(s)}$ of the signal and its transform, one can obtain the following 1-D and 2-D (for rectangular sampling grid) discrete representations of integral Kirchhoff-Rayleigh-Sommerfeld transform:

$$\alpha_r = \sum_{k=0}^{N-1}a_k \, DKRS^{(1D)}\left(\tilde{k} - \tilde{r}\right) \tag{2.4.1}$$

and

$$\alpha_{r,s} = \sum_{k=0}^{N-1}\sum_{l=0}^{N-1} a_{k,l} DKRS^{(2D)}\left(\tilde{k}-\tilde{r};\tilde{l}-\tilde{r}\right) \qquad (2.4.2)$$

where it is denoted:

$$DKRS^{(1D)}_{\tilde{z},\mu}(n) = \frac{\exp\left[i2\pi\dfrac{\tilde{z}^2\sqrt{1+n^2/\tilde{z}^2}}{\mu^2 N}\right]}{1+n^2/\tilde{z}^2}; \qquad (2.4.3,\ a)$$

$$DKRS^{(2D)}_{\tilde{z},\mu}(m,n) = \frac{\exp\left[i2\pi\dfrac{\tilde{z}^2\sqrt{1+m^2/\tilde{z}^2+n^2/\tilde{z}^2}}{\mu^2 N}\right]}{1+m^2/\tilde{z}^2+n^2/\tilde{z}^2}; \qquad (2.4.3,\ b)$$

$$\tilde{z} = Z/\Delta f\ ; \qquad \mu^2 = \frac{\lambda Z}{N\Delta f^2} = \frac{\lambda\tilde{z}}{N\Delta f} \qquad (2.4.3,\ c)$$

We refer to these transforms as 1-D and 2-D **Discrete Kirchhoff-Rayleigh-Sommerfeld Transforms (DKRST)**. When $\tilde{z}\to 0$, DKRS-transform degenerates into the identical transform because object and hologram planes coincide. When $\tilde{z}\to\infty$, DKRS-transform converts to discrete Fresnel transforms. In distinction from 2-D DFTs and DFrTs, 2-D Discrete KRS-transform is an inseparable transform.

As one can see from Eqs. 2.4.3, Discrete Kirchhoff-Rayleigh-Sommerfeld transforms are digital convolutions. Therefore, they can be computed using fast convolution algorithms based on Fourier or fast DCT transforms.

**Appendix**

**A2.1. Computation of convolution of signals with "mirror reflection" extension using Discrete Cosine and Discrete Cosine/Sine Transforms.**

Let signal $\{\tilde{a}_k\}$ be obtained from signal $\{a_k\}$ of $N$ samples by its extension by "mirror reflection" and periodical replication of the result with a period of $2N$ samples:

$$\tilde{a}_{(k)\bmod 2N} = \begin{cases} a_k, & k=0,1,...,N-1; \\ a_{2N-k-1}, & k=N,N=1,...,2N-1 \end{cases} \qquad (A2.1.1)$$

and let $\{h_n\}$ be a convolution kernel of $N$ samples ($n=0,1,...,N-1$). Then

$$\tilde{c}_k = \sum_{n=0}^{N-1} h_n \tilde{a}_{\left(k-n+\left[N/2\right]\right)\bmod 2N} \qquad (A2.1.2)$$

where

$$\left[\frac{N}{2}\right] = \begin{cases} N/2, & \text{for even } N \\ (N\text{-}1)/2, & \text{for odd } N \end{cases} \tag{A2.1.3}$$

is a convolution of the extended signal $\{\tilde{a}_k\}$ with kernel $\{h_n\}$. It will coincide with the cyclic convolution of period $2N$:

$$c_{(k)\bmod 2N} = \sum_{n=0}^{N-1} \tilde{h}_{(n)\bmod 2N}\, \tilde{a}_{\left(k-n+\left[N/2\right]\right)\bmod 2N} \tag{A2.1.4}$$

for kernel

$$\tilde{h}_{(n)\bmod 2N} = \begin{cases} 0, & n = 0,\dots,\left[N/2\right]-1 \\ h_{n-\left[N/2\right]}, & n = \left[N/2\right],\dots,\left[N/2\right]+N-1 \\ 0, & \left[N/2\right]+N,\dots,2N-1 \end{cases}. \tag{A2.1.5}$$

The cyclic convolution described by Eqs. A2.1.4 and 5 is illustrated in Fig. A2.1.

Consider computing convolution by means of IDFT of product of DFT spectra of signals to be convolved. DFT spectrum of the extended signal $\{\tilde{a}_k\}$ is:

$$\tilde{\alpha}_r = \frac{1}{\sqrt{2N}} \sum_{k=0}^{2N-1} \tilde{a}_k \exp\left(i2\pi \frac{kr}{2N}\right) =$$

$$\left\{ \frac{2}{\sqrt{2N}} \sum_{k=0}^{N-1} a_k \cos\left[\pi \frac{(k+1/2)r}{N}\right] \right\} \exp\left(-i\pi \frac{r}{2N}\right) = \alpha_r^{(DCT)} \exp\left(-i\pi \frac{r}{2N}\right),$$

$$\tag{A2.1.6}$$

where

$$\alpha_r^{(DCT)} = \mathrm{DCT}\{a_k\} = \frac{2}{\sqrt{2N}} \sum_{k=0}^{N-1} a_k \cos\left(\pi \frac{k+1/2}{N}\right) \tag{A2.1.7}$$

is Discrete Cosine Transform (DCT) of the initial signal $\{a_k\}$. Therefore, DFT spectrum of the signal extended by "mirror reflection" can be computed via DCT using Fast DCT algorithm. From properties of DCT it follows that DCT spectra feature the following symmetry property:

$$\alpha_N^{DCT} = 0\,;\ \alpha_k^{DCT} = -\alpha_{2N-k}^{DCT} \tag{A2.1.8}$$

$$E\ x\ t\ e\ n\ d\ e\ d \quad s\ i\ g\ n\ a\ l \ \left\{\tilde{a}_k\right\}$$

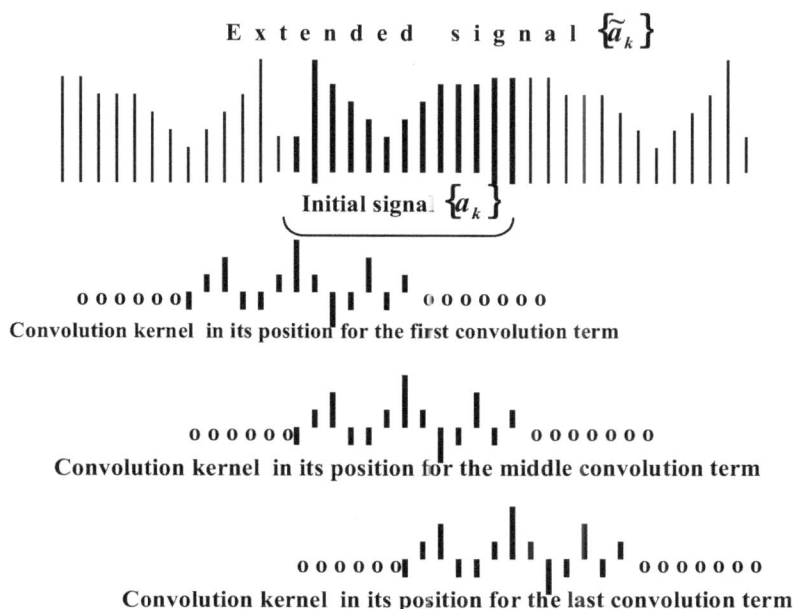

**Fig.A2.1.** Cyclic convolution of a signal extended by its mirror reflection from its borders

Let $\left\{\tilde{\eta}_r\right\}$ be the DFT spectral coefficients of samples of the convolution kernel $\left\{\tilde{h}_n\right\}$, $n = 0, 1, ..., N-1$:

$$\tilde{\eta}_r = \frac{1}{\sqrt{2N}} \sum_{n=0}^{2N-1} \tilde{h}_{(n)\bmod 2N} \exp\left(i2\pi \frac{nr}{2N}\right). \tag{A2.1.9}$$

As $\left\{\tilde{h}_n\right\}$ are real numbers and, therefore, $\left\{\tilde{\eta}_r\right\}$ feature the symmetry property:

$$\left\{\eta_r = \eta_{2N-r}^*\right\}, \tag{A2.1.10}$$

where asterisk symbolizes complex conjugate. For computing convolution, the signal spectrum $\left\{\tilde{\alpha}_r\right\}$ defined by Eq. (A2.1.6) should be multiplied by $\left\{\tilde{\eta}_r\right\}$ and then the inverse DFT should be computed for the first $N$ samples:

$$b_k = \frac{1}{\sqrt{2N}} \sum_{r=0}^{2N-1} \alpha_r^{(DCT)} \exp\left(-i\pi \frac{r}{2N}\right) \tilde{\eta}_r \exp\left(-i2\pi \frac{kr}{2N}\right), \tag{A2.1.11}$$

From Eqs. (A2.1.9) and (A2.1.10), one can now obtain that:

$$b_k = \frac{1}{\sqrt{2N}} \left\{ \alpha_0^{(DCT)} \eta_0 + \sum_{r=1}^{N-1} \alpha_r^{(DCT)} \eta_r \exp\left(-i2\pi \frac{k+1/2}{2N} r\right) + \right.$$

$$\left. + \sum_{r=1}^{N-1} \alpha_{2N-r}^{(DCT)} \eta_{2N-r} \exp\left[-i2\pi \frac{k+1/2}{2N}(2N-r)\right] \right\} =$$

$$\frac{1}{\sqrt{2N}}\left\{\alpha_0^{(DCT)}\eta_0 + \sum_{r=1}^{N-1}\alpha_r^{(DCT)}\left[\eta_r\exp\left(-i2\pi\frac{k+1/2}{2N}r\right) + \eta_r^*\exp\left(i2\pi\frac{k+1/2}{2N}r\right)\right]\right]$$

$$\text{(A2.1.12)}$$

As

$$\eta_r\exp\left(-i2\pi\frac{k+1/2}{2N}r\right) + \eta_r^*\exp\left(i2\pi\frac{k+1/2}{2N}r\right) =$$

$$2\,\mathrm{Re}\left[\eta_r\exp\left(-i2\pi\frac{k+1/2}{2N}r\right)\right] = \eta_r^{re}\cos\left(\pi\frac{k+1/2}{N}r\right) - \eta_r^{im}\sin\left(\pi\frac{k+1/2}{N}r\right),$$

$$\text{(A2.1.13)}$$

where $\{\eta_r^{re}\}$ and $\{\eta_r^{im}\}$ are real and imaginary parts of $\{\eta_r\}$, we obtain:

$$b_k = \frac{1}{\sqrt{2N}}\left\{\alpha_0^{(DCT)}\eta_0 + \sum_{r=1}^{N-1}\alpha_r^{(DCT)}\eta_r^{re}\cos\left(\pi\frac{k+1/2}{N}r\right) - \sum_{r=1}^{N-1}\alpha_r^{(DCT)}\eta_r^{im}\sin\left(\pi\frac{k+1/2}{N}r\right)\right\}.$$

$$\text{(A2.1.14)}$$

First two terms of this expression constitute inverse DCT of the product $\left\{\alpha_r^{(DCT)}\eta_r^{re}\right\}$ while the third term is Discrete Cosine/Sine Transform (DcST) of the product $\left\{\alpha_r^{(DCT)}\eta_r^{im}\right\}$. This expression can be converted to two DCTs:

$$b_k = \frac{1}{\sqrt{2N}}\left\{\alpha_0^{(DCT)}\eta_0 + \sum_{r=1}^{N-1}\alpha_r^{(DCT)}\eta_r^{re}\cos\left(\pi\frac{k+1/2}{N}r\right) - \right.$$

$$\left.\sum_{r=1}^{N-1}\alpha_{N-r}^{(DCT)}\eta_{N-r}^{im}\sin\left(\pi\frac{(k+1/2)(N-r)}{N}\right)\right\} =$$

$$\frac{1}{\sqrt{2N}}\left\{\alpha_0^{(DCT)}\eta_0 + \sum_{r=1}^{N-1}\alpha_r^{(DCT)}\eta_r^{re}\cos\left(\pi\frac{k+1/2}{N}r\right) - \right.$$

$$\left.\sum_{r=1}^{N-1}\alpha_{N-r}^{(DCT)}\eta_{N-r}^{im}\left\{\sin\left[\pi(k+1/2)\right]\cos\left(\pi\frac{k+1/2}{N}r\right) - \cos\left[\pi(k+1/2)\right]\sin\left[\pi\frac{(k+1/2)r}{N}\right]\right\}\right\}$$

$$=\frac{1}{\sqrt{2N}}\left\{\alpha_0^{(DCT)}\eta_0 + \sum_{r=1}^{N-1}\alpha_r^{(DCT)}\eta_r^{re}\cos\left(\pi\frac{k+1/2}{N}r\right) - \right.$$

$$\left.(-1)^k\sum_{r=1}^{N-1}\alpha_{N-r}^{(DCT)}\eta_{N-r}^{im}\cos\left(\pi\frac{k+1/2}{N}r\right)\right\} \qquad \text{(A2.1.15)}$$

## A2.2. Inverse scaled DFT

Scaled DFT

$$\alpha_r^{\sigma} = \frac{1}{\sqrt{\sigma N}} \sum_{k=0}^{N-1} a_k \exp\left( i2\pi \frac{kr}{\sigma N} \right) \qquad (A2.2.1)$$

has its inverse only if $\sigma N$ is an integer number ($\sigma N \in \mathbb{Æ}$). In this case, inverse ScDFT, for $\sigma > 1$, is defined as:

$$a_k^{\sigma} = \frac{1}{\sqrt{\sigma N}} \sum_{r=0}^{\sigma N-1} \alpha_r \exp\left( -i2\pi \frac{kr}{\sigma N} \right) = \begin{cases} a_k, & k = 0,1,...N-1 \\ 0, & k = N, N+1, ..., \sigma N - 1 \end{cases}$$

Proof:

$$a_k^{\sigma} = \frac{1}{\sqrt{\sigma N}} \sum_{r=0}^{\sigma N-1} \alpha_r \exp\left( -i2\pi \frac{kr}{\sigma N} \right) = \frac{1}{\sigma N} \sum_{r=0}^{\sigma N-1} \left[ \sum_{n=0}^{N-1} a_k \exp\left( i2\pi \frac{nr}{\sigma N} \right) \right] \exp\left( -i2\pi \frac{kr}{\sigma N} \right) =$$

$$\frac{1}{\sigma N} \sum_{n=0}^{N-1} a_k \left[ \sum_{r=0}^{\sigma N-1} \exp\left( i2\pi \frac{n-k}{\sigma N} r \right) \right] = \frac{1}{\sigma N} \sum_{n=0}^{N-1} a_k \frac{\exp\left[ i2\pi (n-k) \right] - 1}{\exp\left( i2\pi \frac{n-k}{\sigma N} \right) - 1} =$$

$$\frac{1}{\sigma N} \sum_{n=0}^{N-1} a_n \frac{\sin\left[ \pi (n-k) \right]}{\sin\left( \pi \frac{n-k}{\sigma N} \right)} \exp\left[ i\pi \frac{\sigma N - 1}{\sigma N} (n-k) \right] = \begin{cases} a_n, & k = 0,1,...,N-1 \\ 0, & k = N, N=1,...,\sigma N - 1 \end{cases}$$

$$(A2.2.2)$$

as for w $k = N, N = 1,...,\sigma N - 1$ , $\sin\left[ \pi (n-k) \right] = 0$ and $\sin\left( \pi \frac{n-k}{\sigma N} \right) \neq 0$

## A2.3. Approximation of function frincd(.)

Consider a discrete analog of the known relationship ([10]):

$$\int_{-\infty}^{\infty} \exp\left( i\pi\sigma^2 x^2 \right) \exp\left( -i2\pi fx \right) dx = \frac{\sqrt{i}}{\sigma} \exp\left( -i\pi \frac{f^2}{\sigma^2} \right) \qquad (A2.3.1)$$

By definition of integral:

$$\int_{-\infty}^{\infty} \exp\left( i\pi\sigma^2 x^2 \right) \exp\left( -i2\pi fx \right) dx = \lim_{\substack{N \to \infty \\ \Delta x \to 0}} \sum_{k=-N/2}^{N/2-1} \exp\left( i\pi\sigma^2 k^2 \Delta x^2 \right) \exp\left( -i2\pi rk\Delta x\Delta f \right) \Delta x ,$$

$$(A2.3.2)$$

where $x = k\Delta x$, and the integral is considered in points $f = r\Delta f$. Select $\Delta f \Delta x = 1 / N$ and assume that $N$ is an odd number. Then

$$\int_{-\infty}^{\infty} \exp\left( i\pi\sigma^2 x^2 \right) \exp\left( -i2\pi fx \right) dx = \lim_{\substack{N \to \infty \\ \Delta x \to 0}} \sum_{k=-(N-1)/2}^{(N-1)/2} \exp\left( i\pi\sigma^2 k^2 \Delta x^2 \right) \exp\left( -i2\pi rk\Delta x\Delta f \right) \Delta x$$

$$(A2.3.3)$$

Therefore

$$\lim_{\substack{N \to \infty \\ \Delta x \to 0}} \frac{1}{N\Delta f} \sum_{k=-(N-1)/2}^{(N-1)/2} \exp\left( i\pi \frac{\sigma^2}{N\Delta f^2} \frac{k^2}{N^2} \right) \exp\left( -i2\pi \frac{rk}{N} \right) = \lim_{N \to \infty} \frac{\sqrt{i}}{\sigma} \exp\left( -i\pi \frac{r^2 \Delta f^2}{\sigma^2} \right).$$

$$(A2.3.4)$$

Denote: $\sigma^2 / N\Delta f^2 = q$. Then

$$\lim_{N \to \infty} \frac{1}{N\Delta f} \sum_{k=-(N-1)/2}^{(N-1)/2} \exp\left( i\pi \frac{qk^2}{N} \right) \exp\left( -i2\pi \frac{rk}{N} \right) = \lim_{N \to \infty} \frac{\sqrt{i}}{\Delta f \sqrt{Nq}} \exp\left( -i\pi \frac{r^2 \Delta f^2}{q\Delta f^2 N} \right) =$$

$$\lim_{N \to \infty} \frac{\sqrt{i}}{\Delta f \sqrt{Nq}} \exp\left( -i\pi \frac{r^2}{qN} \right),$$

$$(A2.3.5)$$

or

$$\lim_{N \to \infty} \frac{1}{N} \sum_{k=-(N-1)/2}^{(N-1)/2} \exp\left( i\pi \frac{qk^2}{N} \right) \exp\left( -i2\pi \frac{rk}{N} \right) = \lim_{N \to \infty} \frac{\sqrt{i}}{\sqrt{Nq}} \exp\left( -i\pi \frac{r^2}{qN} \right). \qquad (A2.3.6)$$

Therefore, one can say that

$$\frac{1}{N} \sum_{k=-(N-1)/2}^{(N-1)/2} \exp\left( i\pi \frac{qk^2}{N} \right) \exp\left( -i2\pi \frac{rk}{N} \right) \cong \frac{\sqrt{i}}{\sqrt{Nq}} \exp\left( -i\pi \frac{r^2}{qN} \right). \qquad (A2.3.7)$$

It is assumed in this formula that both $k$ and $r$ are running in the range $-(N-1)/2,...,0,...(N-1)/2$. Introduce variables $n = k + (N-1)/2$ and $s = r + (N-1)/2$ to convert this formula to the formula that corresponds to the canonical DFT in which variables $n$ and $s$ run in the range $0,...,N-1$.

$$\frac{1}{N} \sum_{k=-(N-1)/2}^{(N-1)/2} \exp\left( i\pi \frac{qk^2}{N} \right) \exp\left( -i2\pi \frac{rk}{N} \right) =$$

$$\frac{1}{N} \sum_{n=0}^{N-1} \exp\left\{ i\pi \frac{q[n-(N-1)/2]^2}{N} \right\} \exp\left\{ -i2\pi \frac{[s-(N-1)/2][n-(N-1)/2]}{N} \right\} =$$

$$\frac{1}{N} \exp\left[ i\pi \left( \frac{q}{2} - 1 \right) \frac{(N-1)^2}{2N} \right] \exp\left[ i\pi \frac{s(N-1)}{N} \right] \times$$

$$\sum_{n=0}^{N-1} \exp\left( i\pi \frac{qn^2}{N} \right) \exp\left[ i\pi \frac{(1-q)(N-1)}{N} n \right] \exp\left( -i2\pi \frac{ns}{N} \right) \cong$$

$$\frac{\sqrt{i}}{\sqrt{Nq}} \exp\left\{ -i\pi \frac{[s-(N-1)/2]^2}{qN} \right\}. \qquad (A2.3.8)$$

From this we have:

$$\frac{1}{N}\sum_{n=0}^{N-1}\exp\left(i\pi\frac{qn^2}{N}\right)\exp\left(-i2\pi\frac{n\left[s-\left(1-q\right)(N-1)/2\right]}{N}\right)\cong$$

$$\frac{\sqrt{i}}{\sqrt{Nq}}\exp\left\{-i\pi\left[\left(\frac{q}{2}-1\right)\frac{\left(N-1\right)^2}{2N}\right]\right\}\exp\left[-i\pi\frac{s\left(N-1\right)}{N}\right]\exp\left\{-i\pi\frac{\left[s-\left(N-1\right)/2\right]^2}{qN}\right\}$$

$$(A2.3.9)$$

within boundaries $\left(1-q\right)\dfrac{N-1}{2}<s<\left(1+q\right)\dfrac{N-1}{2}$; $0\le q\le 1$, or finally

$$\frac{1}{N}\sum_{n=0}^{N-1}\exp\left(i\pi\frac{qn^2}{N}\right)\exp\left[-i2\pi\frac{n\left(s-v_q\right)}{N}\right]\cong\sqrt{\frac{i}{Nq}}\exp\left[-i\pi\frac{\left(s-v_q\right)^2}{qN}\right]\mathrm{rect}\left[\frac{s-v_q}{q\left(N-1\right)}\right],$$

$$(A2.3.10)$$

where $v_q=\left(1-q\right)(N-1)/2$ , $0\le q\le 1$ and $s$ runs from $0$ to $N-1$.
Numerical evaluation of the relationship

$$\mathrm{RCT}\left(N;q;s\right)=\frac{1}{N}\sum_{n=0}^{N-1}\exp\left(i\pi\frac{qn^2}{N}\right)\exp\left[-i2\pi\frac{n\left(s-v_q\right)}{N}\right]\Bigg/\sqrt{\frac{i}{Nq}}\exp\left[-i\pi\frac{\left(s-v_q\right)^2}{qN}\right]=$$

$$\frac{\sqrt{q}\exp\left[i\pi\frac{\left(s-v_q\right)^2}{qN}\right]}{\sqrt{iN}}\sum_{n=0}^{N-1}\exp\left(i\pi\frac{qn^2}{N}\right)\exp\left[-i2\pi\frac{n\left(s-v_q\right)}{N}\right]\cong\mathrm{rect}\left[\frac{s-v_q}{q\left(N-1\right)}\right]$$

$$(A2.3.11)$$

confirms the validity of this approximation. This is illustrated in Fig. A2.2 for $N$=512 and four different values of $q$ (0.25; 0.5; 0.75 and 1.0). Rect-function of Eq. A2.3.11 is shown in bold line.

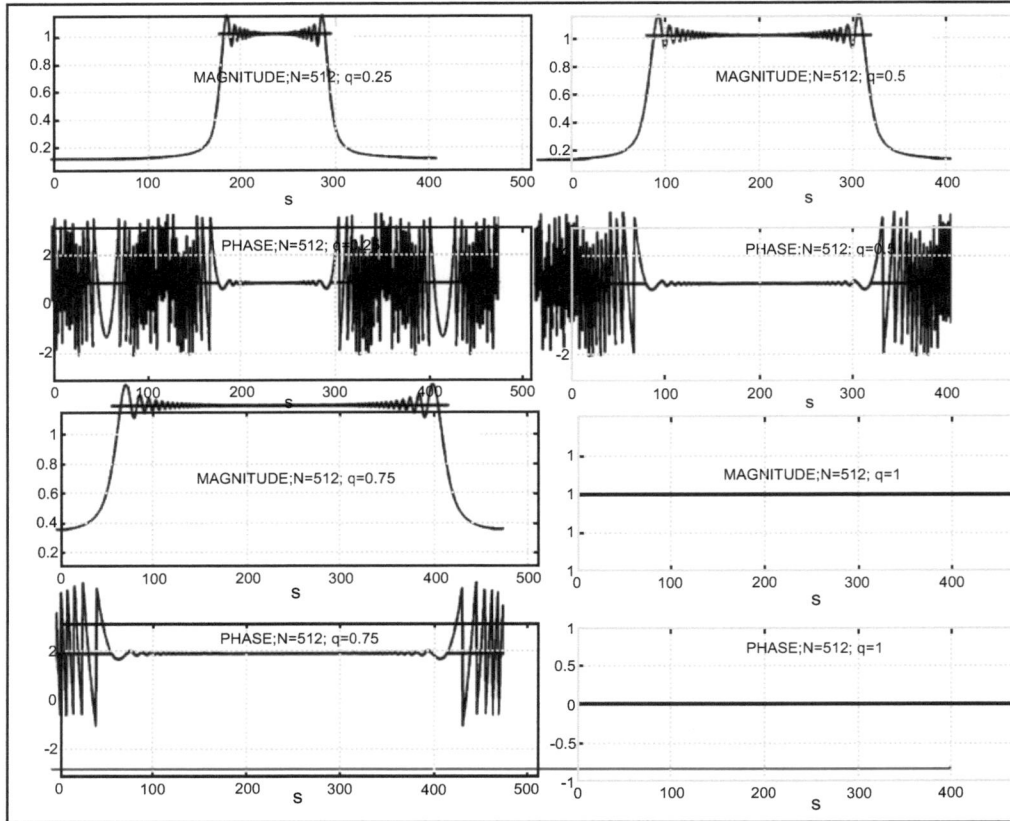

**Fig. A2.2.** Plots of absolute values (magnitude) and phase (in radians) of function RCT(N;q,s, Eq. A2.3.11, for N=512 and four different values of q (0.25; 0.5; 0.75 and 1.0). Rect-function that approximates this function is shown in bold line.

## 2.5    A2.4. Summary of the fast discrete diffraction transforms
Table.

| Transform | |
|---|---|
| Canonical Discrete Fourier Transform (DFT) | $\alpha_r = \dfrac{1}{\sqrt{N}} \sum_{k=0}^{N-1} a_k \exp\left( i2\pi \dfrac{kr}{N} \right)$ |
| Shifted DFT | $\alpha_r^{u,v} = \dfrac{1}{\sqrt{N}} \sum_{k=0}^{N-1} a_k \exp\left[ i2\pi \dfrac{(k+u)(r+v)}{N} \right] = \dfrac{1}{\sqrt{N}} \sum_{k=0}^{N-1} a_k \exp\left( i2\pi \dfrac{\tilde{k}\tilde{r}}{N} \right)$ |
| Discrete Cosine Transform (DCT) | $\alpha_r^{DCT} = \dfrac{2}{\sqrt{2N}} \sum_{k=0}^{N-1} a_k \cos\left( \pi \dfrac{k+1/2}{N} r \right)$ |
| Discrete Cosine-Sine Transform (DcST) | $\alpha_r^{DcST} = \dfrac{2}{\sqrt{2N}} \sum_{k=0}^{N-1} a_k \sin\left( \pi \dfrac{k+1/2}{N} r \right)$ |

| Transform | |
|---|---|
| Scaled DFT | $$\alpha_r^\sigma = \frac{1}{\sqrt{\sigma N}} \sum_{k=0}^{N-1} a_k \exp\left[ i2\pi \frac{(k+u)(r+v)}{\sigma N} \right] = \frac{1}{\sqrt{\sigma N}} \sum_{k=0}^{N-1} a_k \exp\left( i2\pi \frac{\tilde{k}\tilde{r}}{\sigma N} \right)$$ |
| Scaled DFT as a cyclic convolution | $$\alpha_r^\sigma = \frac{\exp\left( i\pi \frac{\tilde{r}^2}{\sigma N} \right)}{\sqrt{\sigma N}} \sum_{k=0}^{N-1} \left[ a_k \exp\left( i\pi \frac{\tilde{k}^2}{\sigma N} \right) \right] \exp\left[ -i\pi \frac{(\tilde{k}-\tilde{r})^2}{\sigma N} \right]$$ |
| Canonical 2-D DFT | $$\alpha_{r,s} = \frac{1}{\sqrt{N_1 N_2}} \sum_{k=0}^{N_1-1} \sum_{l=0}^{N_2-1} a_{k,l} \exp\left[ i2\pi \left( \frac{kr}{N_1} + \frac{ls}{N_1} \right) \right]$$ |
| Affine DFT | $$\alpha_{r,s} = \sum_{k=0}^{N_1-1} \sum_{l=0}^{N_2-1} a_{k,l} \exp\left[ i2\pi \left( \frac{rk}{\sigma_A N_1} + \frac{sk}{\sigma_C N_1} + \frac{rl}{\sigma_B N_2} + \frac{sl}{\sigma_D N_2} \right) \right]$$ |
| Rotated DFT (RotDFT) | $$\alpha_{r,s} = \sum_{k=0}^{N-1} \sum_{l=0}^{N-1} a_{k,l} \exp\left[ i2\pi \left( \frac{r\cos\theta - s\sin\theta}{N} k + \frac{r\sin\theta + s\cos\theta}{N} l \right) \right] =$$ $$\sum_{k=0}^{N-1} \sum_{l=0}^{N-1} a_{k,l} \exp\left[ i2\pi \left( \frac{rk+sl}{N}\cos\theta - \frac{sk-rl}{N}\sin\theta \right) \right]$$ |
| Rotated Scaled DFT | $$\alpha_{r,s} = \sum_{k=0}^{N-1} \sum_{l=0}^{N-1} a_{k,l} \exp\left[ i2\pi \left( \frac{r\cos\theta - s\sin\theta}{\sigma N} k + \frac{r\sin\theta + s\cos\theta}{\sigma N} l \right) \right] =$$ $$\sum_{k=0}^{N-1} \sum_{l=0}^{N-1} a_{k,l} \exp\left[ i2\pi \left( \frac{rk+sl}{\sigma N}\cos\theta - \frac{sk-rl}{\sigma N}\sin\theta \right) \right]$$ |
| Discrete Sinc-function | $$\mathrm{sincd}(N,x) = \frac{\sin x}{N \sin(x/N)}$$ |
| Canonical Discrete Fresnel Transform (DFrT) | $$\alpha_r = \frac{1}{\sqrt{N}} \sum_{k=0}^{N-1} a_k \exp\left[ i\pi \frac{(k/\mu - r\mu)^2}{N} \right]; \qquad \mu^2 = \lambda Z / N \Delta f^2$$ |
| Shifted DFrT | $$\alpha_r^{(\mu,w)} = \frac{1}{\sqrt{N}} \sum_{k=0}^{N-1} a_k \exp\left[ -i\pi \frac{(k\mu - r/\mu + w)^2}{N} \right]; \quad w = u/\mu - v\mu$$ |
| Fourier Reconstruction algorithm for Fresnel holograms | $$\alpha_r^{(\mu,w)} = \frac{\exp\left( -i\pi \frac{r^2}{\mu^2 N} \right)}{\sqrt{N}} \sum_{k=0}^{N-1} a_k \exp\left[ -i\pi \frac{(k\mu + w)^2}{N} \right] \exp\left( i2\pi \frac{k+w/\mu}{N} r \right)$$ |
| Focal Plane invariant DFrT | $$\alpha_r^{\left( \mu, \frac{N}{2\mu} \right)} = \frac{1}{\sqrt{N}} \sum_{k=0}^{N-1} a_k \exp\left\{ -i\pi \frac{[k\mu - (r - N/2)/\mu]^2}{N} \right\}$$ |
| Partial DFrT (PDFT) | $$\widehat{\alpha}_r^{(\mu,w)} = \frac{1}{\sqrt{N}} \sum_{k=0}^{N-1} a_k \exp\left( -i\pi \frac{k^2 \mu^2}{N} \right) \exp\left[ i2\pi \frac{k(r - w\mu)}{N} \right]$$ |

| Convolutional Discrete Fresnel Transform (ConvDFrT) | $\alpha_r = \sum_{k=0}^{N-1} a_k \, \mathrm{frincd}\left(N;\mu^2;r+w-k\right) =$ $\frac{1}{N}\sum_{s=0}^{N-1}\left[\sum_{k=0}^{N-1}a_k\exp\left(i2\pi\frac{k-r-w}{N}s\right)\right]\exp\left(-i\pi\frac{\mu^2 s^2}{N}\right)$ |
|---|---|
| Convolutional reconstruction algorithm for Fresnel holograms | $\alpha_r = \frac{1}{N}\sum_{s=0}^{N-1}\left[\sum_{k=0}^{N-1}a_k\exp\left(i2\pi\frac{ks}{N}\right)\right]\exp\left(-i\pi\frac{\mu^2 s^2}{N}\right)\exp\left(-i2\pi\frac{r+w}{N}s\right)$ |
| Frincd-function | $\mathrm{frincd}\left(N;q;x\right)=\frac{1}{N}\sum_{r=0}^{N-1}\exp\left(i\pi\frac{qr^2}{N}\right)\exp\left(-i2\pi\frac{xr}{N}\right)$ $\mathrm{frincd}\left(N;1;x\right)=\sqrt{\frac{i}{N}}\exp\left(-i\pi\frac{x^2}{N}\right);\ x\in\mathbb{Z}$ Analytical approximation: $\mathrm{frincd}\left(N;\pm q;x\right)\cong\sqrt{\frac{\pm i}{Nq}}\exp\left[\mp i\pi\frac{x^2}{qN}\right]\mathrm{rect}\left[\frac{x}{q(N-1)}\right]$ |
| Discrete Kirchhoff-Rayleigh-Sommerfeld Transform (DKRST) | $\alpha_r=\sum_{k=0}^{N-1}a_k\dfrac{\exp\left[i2\pi\dfrac{\tilde{z}^2\sqrt{1+\left(\tilde{k}-\tilde{r}\right)^2\big/\tilde{z}^2}}{\mu^2 N}\right]}{1+\left(\tilde{k}-\tilde{r}\right)^2\big/\tilde{z}^2}$ |

## 2.6   References

1. L. Yaroslavsky, Digital Holography and Digital Image Processing, Kluwer Academic Publishers, Boston, 2004
2. L. Yaroslavsky, Discrete Transforms, Fast Algorithms and Point Spread Functions of Numerical Reconstruction of Digitally Recorded Holograms, In: Advances in Signal Transforms: Theory and Applications, J. Astola and L. Yaroslavsky, Eds., Hindawi Publishing Corporation, 2007
3. L. R. Rabiner, R. W. Schafer, C. M. Rader, The chirp z-transform algorithm and its application, Bell System Tech. J., 1969, Vol. 48: 1249-1292
4. Xuegong Deng, Bipin Bihari, Jianhua Gan, Feng Zhao, and Ray T. Chen, Fast algorithms for chirp transforms with zooming-in ability and its application, J. Opt. Soc. Am. A, Vol. 17, No. 4, April 2000
5. V. Namias, The Fractional Order Fourier Transform and its Applications to quantum mechanics, J. Inst. Math. Appl. v. 25, 241-265 (1980)
6. D. H. Bailey, P. N. Scwarztrauber, The Fractional Fourier Transform and applications, SIAM Rev. 1991, Vol. 33, 389-404
7. L. Yaroslavsky and N. Ben David, Focal plane invariant algorithm for digital reconstruction of holograms recorded in the near diffraction zone, In: Optical Measurement Systems for Industrial Inspection III, SPIE's Int. Symposium on Optical Metrology, 23-25 June 2003, Munich, Germany, W. Osten, K. Creath, M. Kujawinska, Eds., SPIE v. 5144, pp. 142-149
8. G. Pedrini, H. J. Tiziani, "Short-Coherence Digital Microscopy by Use of a Lensless Holographic Imaging System", Applied Optics-OT, 41, (22), 4489-4496, (2002).
9. E. Ciche, P. Marquet, Chr. Depeursinge, "Spatial filtering of zero-order and twin-image elimination in digital off-axis holography, Appl. Optics, v. 30, No. 23, Aug. 2000
10. I. S. Gradstein, I. M. Ryzhik, "Tables of integrals, series, and products", Academic Press, 1994.

## CHAPTER 3

## 3   Digital Recording and Numerical Reconstruction of Holograms

**Abstract:** Digital recording and numerical reconstruction of holograms is one of the main tasks of digital holography with numerous applications. This chapter exposes off-axis and on-axis methods of hologram digital recording, introduces the notion of the point-spread function of numerical reconstruction of hologram, which is the main metrological characteristics of the process, presents results of evaluation of point-spread functions of numerical reconstruction of Fourier and Fresnel holograms and discusses zones of applicability of different reconstruction algorithms

### 3.1   Introduction

First experiments in numerical reconstruction of optical holograms date back to 1960-70-th ([1-3]). At that time, scanning devices that could be used for digitizing holograms, had low resolution, which required optical magnification of holograms to fit them to the resolution of scanning devices. Fig. 3.1 reproduces the result reported in [1, 2].

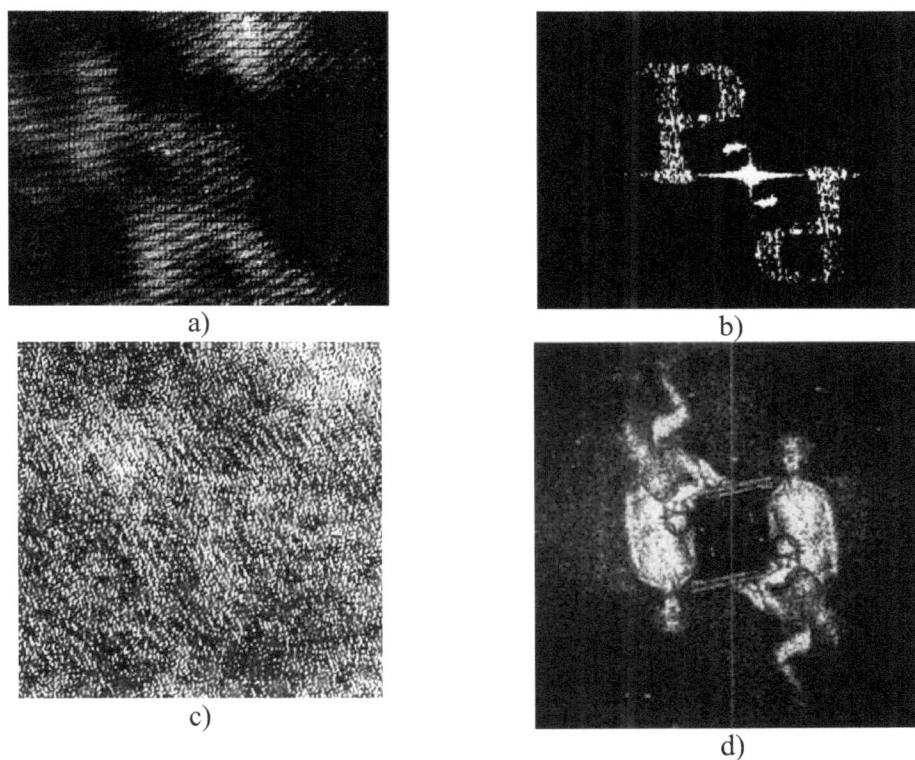

a)

b)

c)

d)

**Fig. 3.1.** First holograms numerically reconstructed in computers: a) a test optical Fourier hologram electronically recorded using vidicon TV camera to 256x256 pixels quantized to 8 grey levels; b) – image numerically reconstructed from this hologram on a computer PDP-6; c) - a test optical Fourier hologram optically magnified with magnification factor 20, printed and scanned to 512x512 pixels using electro-mechanical scanner with resolution 0.2 mm and quantized to 64 gray levels in a logarithmic scale; d) - two conjugate images reconstructed from this hologram on computer Minsk-22[1]. Images a) and b) are adopted from Ref. 1., images c) and d) are adopted from Ref. 2

---

[1] http://www.computer-museum.ru/english/minsk0.htm

Then, with an advent of digital CCD cameras with pitch of 10-20 mcm, it had become possible to perform direct digitizing optical hologram in the process of hologram recording. In first experiments with CCD cameras ([4]), holograms were recorded in the Leith-Upatnieks off axis scheme ([5]). Later, phase-shifting method of recording holograms was suggested ([6]) that enabled on-axis hologram recording scheme more efficient in terms of the use of resolution of digital cameras. Since that time, numerous projects in digital recording and numerical reconstruction of optical holograms have been initiated and implemented, especially in the field of optical holographic microscopy.

In this chapter, mathematical models of off-axis and on-axis methods for digital recording of holograms are introduced in Sects. 3.2 and 3.3. In Sect. 3.4, point spread functions of numerical reconstruction of holograms are derived. In Sect. 3.5, zones of applicability of Fourier and Convolution algorithms for numerical reconstruction of Fresnel holograms are discussed.

## 3.2   "Off-axis" digital recording of holograms

Schematic diagram of "off axis" method for electronic recording of holograms is presented in Fig. 3.2. In the off-axis recording, a spatial angle between the reference and object beams is introduced that exceeds the angular size of the object, under which it is seen from the hologram plane.

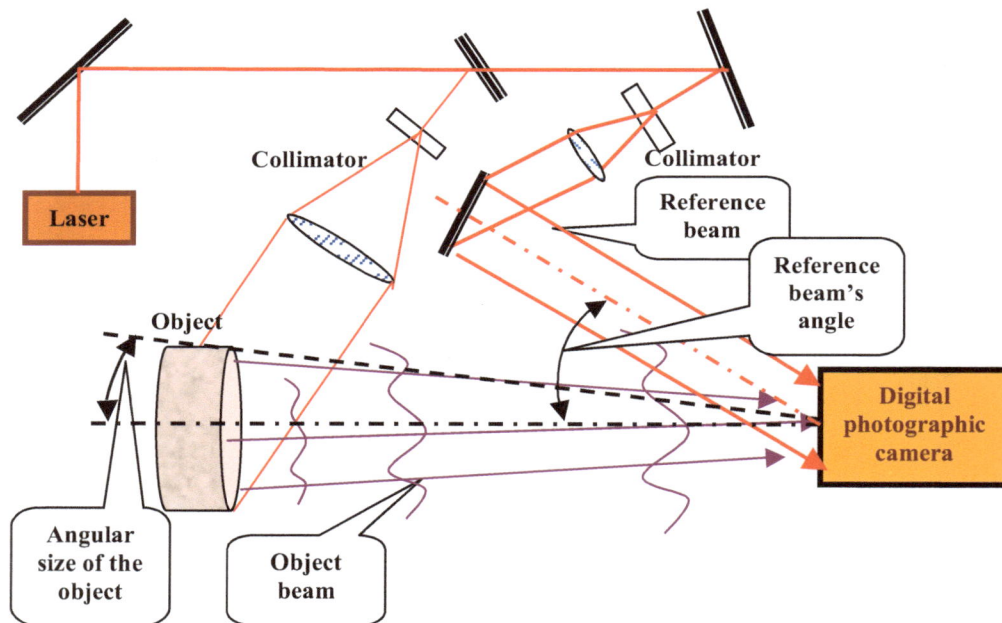

**Fig. 3.1.** Schematic diagram of the "off-axis" method for electronic hologram recording.

Consider the mathematical model of recording holograms

$$\mathrm{H}\left(f_x,f_y\right)=\left|\alpha\left(f_x,f_y\right)+R\left(f_x,f_y\right)\right|^2=$$

$$\alpha\left(f_x,f_y\right)R^*\left(f_x,f_y\right)+\alpha^*\left(f_x,f_y\right)R\left(f_x,f_y\right)+\left|\alpha\left(f_x,f_y\right)\right|^2+\left|R\left(f_x,f_y\right)\right|^2, \quad (3.2.1)$$

described in Ch. 1 (Eq. 1.2.2). Here $\alpha\left(f_x, f_y\right)$ and $R\left(f_x, f_y\right)$ denote complex amplitudes of the object and reference beams, respectively, at point $\left(f_x, f_y\right)$ of the hologram plane and asterisk denotes complex conjugate. Numerical reconstruction of the hologram consists in applying to the recorded hologram a discrete transform that implements wave propagation from the hologram plane back to object.

Eq. 3.2.1 can be modified to a form:

$$H\left(f_x, f_y\right) = \left|\alpha\left(f_x, f_y\right)\right|^2 + \left|R\left(f_x, f_y\right)\right|^2$$
$$+ 2\left|\alpha\left(f_x, f_y\right)\right|\left|R\left(f_x, f_y\right)\right|\cos\left[\theta_{Obj}\left(f_x, f_y\right) - \theta_{Re\,j}\left(f_x, f_y\right)\right], \qquad (3.2.2)$$

which explicitly shows that the recorded hologram signal contains a term with an amplitude and phase modulated spatial carrier (the last term in Eq. 3.2.2). This spatial carrier forms periodical patterns that can be quite clearly seen in the holograms shown in Figs. 3.1 a) and c). It is also seen in the results of hologram reconstruction shown in Figs. 3.1, b) and d) that the first term in Eq. 3.2.2 generates the central bright spot in the reconstructed images, or so-called **zero-order diffraction term**, while the second term, the spatial carrier one, produces two images, the direct and conjugated ones. As one can see, the distance between centers of these images must exceed the image size, otherwise the direct and conjugate images will overlap and can't be separated. Because of this, the highest spatial frequency of the recorded hologram is at least twice as that of the object image. Therefore the hologram recording device must have at least twice as much resolution cells for recording and sampling the holograms compared to that required for recording of only the mathematical hologram (the first term in the right part of Eq. 3.2.1), which will reconstruct the object image without the conjugate image.

## 3.3  "On-axis" digital recording of holograms

Schematic diagram of the "off-axis" method for electronic recording of holograms is shown in Fig. 3.3. In this method, known also as phase-shifting holography ([6]), object and reference beams are co-linear (Fig. 3.2) and several exposures of holograms of the object are carried out with shifting, at each exposure, the phase of the reference beam plane wave front and recording the results of the exposures. The recording results are then combined in computer in a certain way, which depends on the number of exposures and phase shifts of the reference beam in each exposure, to compute from them the mathematical hologram.

In what follows, we show that at least three exposures are required in this method. Let $\theta_k$ be a phase shift of the reference beam in $k$-th hologram exposure, $k = 1, ..., K$. Then

$$H_k\left(f_x, f_y\right) = \left|\alpha\left(f_x, f_y\right) + R\left(f_x, f_y\right)\exp\left(i\theta_k\right)\right|^2 = \left|\alpha\left(f_x, f_y\right)\right|^2 + \left|R\left(f_x, f_y\right)\right|^2 +$$
$$\alpha\left(f_x, f_y\right)R^*\left(f_x, f_y\right)\exp\left(-i\theta_k\right) + \alpha^*\left(f_x, f_y\right)R\left(f_x, f_y\right)\exp\left(i\theta_k\right) \qquad (3.3.1)$$

is a hologram recorded in $k$-th exposure. For separating the mathematical hologram term containing $\alpha\left(f_x, f_y\right)$, $K$ recorded at each exposure holograms $\left\{H_k\right\}$ are summed up with the same phase shift as those used in their recording:

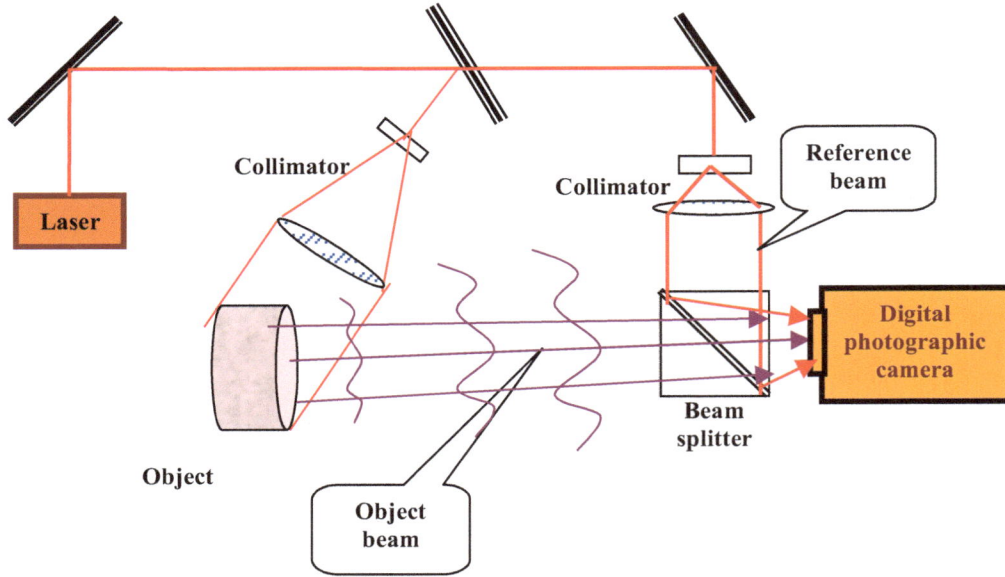

**Fig. 3.2.** Schematic diagram of the "off axis" method for electronic recording of holograms

$$\bar{H} = \frac{1}{K}\sum_{k=1}^{K} H_k \exp\left(i\theta_k\right) = \alpha\left(f_x, f_y\right) R^*\left(f_x, f_y\right) +$$

$$\alpha^*\left(f_x, f_y\right) R\left(f_x, f_y\right)\sum_{k=1}^{K}\exp\left(i2\theta_k\right) + \left[\left|\alpha\left(f_x, f_y\right)\right|^2 + \left|R\left(f_x, f_y\right)\right|\right]\sum_{k=1}^{K}\exp\left(i\theta_k\right). \quad (3.3.2)$$

For eliminating from this sum all terms but the first one, phases $\left\{\theta_k\right\}$ should be solutions of equations:

$$\sum_{k=1}^{K}\exp\left(i\theta_k\right) = 0; \qquad\qquad (3.3.3,\ a)$$

$$\sum_{k=1}^{K}\exp\left(i2\theta_k\right) = 0. \qquad\qquad (3.3.3,\ b)$$

It is convenient to use phase shifts that form an arithmetic progression

$$\theta_k = \left(k-1\right)\theta_0, \quad k = 1,\dots, N \qquad\qquad (3.3.4)$$

Then

$$\sum_{k=1}^{K} \exp\left(i\theta_k\right) = \sum_{k=1}^{K} \exp\left[i(k-1)\theta_0\right] = \frac{\exp\left(iK\theta_0\right)-1}{\exp\left(i\theta_0\right)-1} ;$$

$$\sum_{k=1}^{K} \exp\left(i2\theta_k\right) = \sum_{k=0}^{K-1} \exp\left[i2(k-1)\theta_0\right] = \frac{\exp\left(i2K\theta_0\right)-1}{\exp\left(i2\theta_0\right)-1} = \frac{\exp\left(iK\theta_0\right)-1}{\exp\left(i\theta_0\right)-1}\frac{\exp\left(iK\theta_0\right)+1}{\exp\left(i\theta_0\right)+1} ,$$

$$(3.3.5)$$

from which it follows that solution of Eq. 3.3.3, a) is $\theta_0 = \dfrac{2\pi}{K}$ except the solution $\theta_0 = \pi$ for $K = 2$, which does not satisfy Eq. 3.3.3, b) because

$$\frac{\exp\left(i2\theta_0\right)-1}{\exp\left(i\theta_0\right)-1}\frac{\exp\left(i2\theta_0\right)+1}{\exp\left(i\theta_0\right)+1} = \exp\left(i2\pi\right)+1 = 2 . \tag{3.3.6}$$

Therefore

$$\theta_0 = \frac{2\pi}{K} \tag{3.3.7}$$

for any integer $K \geq 3$.

For example, in four-exposures ($K = 4, \theta_0 = \dfrac{\pi}{2}$) we will have from Eq. 3.3.1 the following:
The first exposure records

$$H_1\left(f_x, f_y\right) = \left|\alpha\left(f_x, f_y\right)\right|^2 + \left|R\left(f_x, f_y\right)\right|^2 + \alpha\left(f_x, f_y\right)R^*\left(f_x, f_y\right) + \alpha^*\left(f_x, f_y\right)R\left(f_x, f_y\right);$$

$$(3.3.6\ a)$$

The second exposure records

$$H_1\left(f_x, f_y\right) = \left|\alpha\left(f_x, f_y\right)\right|^2 + \left|R\left(f_x, f_y\right)\right|^2$$
$$- i\alpha\left(f_x, f_y\right)R^*\left(f_x, f_y\right) + i\alpha^*\left(f_x, f_y\right)R\left(f_x, f_y\right) \tag{3.3.6\ b}$$

Third exposure records

$$H_1\left(f_x, f_y\right) = \left|\alpha\left(f_x, f_y\right)\right|^2 + \left|R\left(f_x, f_y\right)\right|^2 - \alpha\left(f_x, f_y\right)R^*\left(f_x, f_y\right) - \alpha^*\left(f_x, f_y\right)R\left(f_x, f_y\right)$$

$$(3.3.6\ c)$$

Fourth exposure records

$$H_1\left(f_x,f_y\right)=\left|\alpha\left(f_x,f_y\right)\right|^2+\left|R\left(f_x,f_y\right)\right|^2$$
$$+i\alpha\left(f_x,f_y\right)R^*\left(f_x,f_y\right)-i\alpha^*\left(f_x,f_y\right)R\left(f_x,f_y\right).$$

(3.3.6 d)

Then the mathematical hologram can be computed as

$$\alpha\left(f_x,f_y\right)R^*\left(f_x,f_y\right)=\frac{H_1+iH_2-H_3-iH_4}{4}$$

(3.3.7)

## 3.4 Point spread functions of numerical reconstruction of holograms

### 3.4.1. A general formulation

In numerical reconstruction of holograms, samples of the object wave front are reconstructed out of samples of its recorded hologram using the discrete diffraction transforms. This process can be treated as sampling the object wave front by a sampling system that consists of the hologram sampling device and a computer, in which the object wav front samples are numerically reconstructed form the hologram.

Signal sampling is a linear transformation that is fully specified by its point spread function (PSF), which establishes a link between an object signal $a(x)$ and its samples $\{a_k\}$:

$$a_k=\int_X a(x)PSF(x,k)dx,$$

(3.4.1)

According to the sampling theory (see, for instance, [7]), for a given sampling interval $\Delta x$, PSF of the ideal sampling device is a sinc-function:

$$PSF(x,k)=\mathrm{sinc}\left[\pi\left(x-k\Delta x\right)\big/\Delta x\right]=\frac{\sin\left[\pi\left(x-k\Delta x\right)\big/\Delta x\right]}{\pi\left(x-k\Delta x\right)\big/\Delta x}$$

(3.4.2)

Provided that the continuous signal is reconstructed from its samples using, as a reconstruction basis functions, the same sinc-functions, this PSF secures least root mean square signal reconstruction error.

In this section, we consider how point spread functions of different reconstruction algorithms depend on algorithm parameters and on physical parameters of holograms and their sampling devices. For the sake of simplicity, we will consider 1-D holograms and transforms. Corresponding 2-D results are straightforward in the conventional assumption of separability of sampling and transforms.

Let, in numerical reconstruction of holograms, samples $\{a_k\}$ of the object wave front be obtained through a transformation

$$a_k = \sum_{r=0}^{N-1} \alpha_r DRK(r,k).$$

(3.4.3)

of available hologram samples $\{\alpha_r\}$ with a certain discrete reconstruction kernel $DRK(r,k)$ that corresponds to the type of the hologram. Let also hologram samples $\{\alpha_r\}$ are measured by a hologram recording and sampling device as

$$\alpha_r = \int_{-\infty}^{\infty} \alpha(f) \varphi_f^{(s)} \left( f - \tilde{r}^{(s)} \Delta f^{(s)} \right) df$$

(3.4.4)

where $\alpha(f)$ is a hologram signal, $\{\varphi_f^{(s)}(.)\}$ is a point spread function of the hologram sampling device, $\Delta f^{(s)}$ is a hologram sampling interval, $\tilde{r}^{(s)} = r + v^{(s)}$, $r$ is an integer index of hologram samples and $v^{(s)}$ is a shift, in units of the hologram sampling interval, of the hologram sampling grid with respect to the hologram coordinate system; these sampling parameters are analogous to those illustrated, for signal sampling, in Fig. 3.2.

The hologram signal $\alpha(f)$ is linked with object wave front $a(x)$ through a diffraction integral

$$\alpha(f) = \int_{-\infty}^{\infty} a(x) WPK(x,f) dx,$$

(3.4.5)

where $WPK(x,f)$ is a wave propagation kernel. Therefore, one can rewrite Eq. 3.4.4 as:

$$\alpha_r = \int_{-\infty}^{\infty} \left[ \int_{-\infty}^{\infty} a(x) WPK(x,f) dx \right] \varphi_f^{(s)} \left( f - \tilde{r}^{(s)} \Delta f^{(s)} \right) df =$$

$$\int_{-\infty}^{\infty} \int_{-\infty}^{\infty} a(x) WPK(x,f) \varphi_f^{(s)} \left( f - \tilde{r}^{(s)} \Delta f^{(s)} \right) dx \, df =$$

$$\int_{-\infty}^{\infty} a(x) dx \int_{-\infty}^{\infty} WPK(x,f) \varphi_f^{(s)} \left( f - \tilde{r}^{(s)} \Delta f^{(s)} \right) df$$

(3.4.6)

Insert now Eq. 3.3.6 into Eq. 3.3.3 and establish a link between the object wave front $a(x)$ and its samples $\{a_k\}$ reconstructed from the sampled hologram:

$$a_k = \sum_{r=0}^{N-1} \left[ \int_{-\infty}^{\infty} a(x) dx \int_{-\infty}^{\infty} WPK(x,f) \varphi_f^{(s)} \left( f - \tilde{r}^{(s)} \Delta f^{(s)} \right) df \right] DRK(r,k) =$$

$$\int_{-\infty}^{\infty} a(x) dx \left[ \int_{-\infty}^{\infty} WPK(x,f) df \sum_{r=0}^{N-1} DRK(r,k) \varphi_f^{(s)} \left( f - \tilde{r}^{(s)} \Delta f^{(s)} \right) \right] = \int_{-\infty}^{\infty} a(x) PSF(x,k) dx$$

$$(3.4.7)$$

where function

$$OPSF(x,k) = \int_{-\infty}^{\infty} WPK(x,f) df \sum_{r=0}^{N-1} DRK(r,k) \varphi_f^{(s)} \left( f - \tilde{r}^{(s)} \Delta f^{(s)} \right) \qquad (3.4.8)$$

can be treated as *an overall point spread function (OPSF) of numerical reconstruction of holograms*. As one can see form Eq. 7.1.8, it depends on all factors involved in the process of sampling and reconstruction of holograms: wave propagation kernel $WPK(.,.)$, discrete reconstruction kernel $DRK(.,.)$ and point spread function of the hologram sampling device $\varphi_f^{(s)}(.)$.

For further analysis, it is convenient to replace point spread function of the hologram sampling device through its Fourier Transform, or its frequency response $\boldsymbol{\Phi}_f^{(s)}(.)$:

$$\varphi_f^{(s)} \left( f - \tilde{r}^{(s)} \Delta f \right) =$$

$$\int_{-\infty}^{\infty} \Phi_f^{(s)}(\xi) \exp\left[ i2\pi \left( f - \tilde{r}^{(s)} \Delta f \right) \xi \right] d\xi = \int_{-\infty}^{\infty} \Phi_f^{(s)}(\xi) \exp\left( -i2\pi \tilde{r}^{(s)} \Delta f^{(s)} \xi \right) \exp\left( i2\pi f \xi \right) d\xi$$

$$(3.4.9)$$

Then obtain:

$$PSF(x,k) = \int_{-\infty}^{\infty} WPK(x,f) df \sum_{r=0}^{N-1} DRK(r,k) \int_{-\infty}^{\infty} \Phi_f^{(s)}(\xi) \exp\left( -i2\pi \tilde{r}^{(s)} \Delta f^{(s)} \xi \right) \exp\left( i2\pi f \xi \right) d\xi =$$

$$\int_{-\infty}^{\infty} \Phi_f^{(s)}(\xi) d\xi \int_{-\infty}^{\infty} WPK(x,f) \exp\left( i2\pi f \xi \right) df \sum_{r=0}^{N-1} DRK(r,k) \exp\left( -i2\pi \tilde{r}^{(s)} \Delta f^{(s)} \xi \right). \quad (3.4.10)$$

Introduce function

$$\overline{PSF}(x,\xi;k) = \int_{-\infty}^{\infty} WPK(x,f) \exp\left( i2\pi f \xi \right) df \sum_{r=0}^{N-1} DRK(r,k) \exp\left( -i2\pi \tilde{r}^{(s)} \Delta f^{(s)} \xi \right) =$$

$$\overline{WPK}(x,\xi) \cdot \overline{DRK}(\xi,k), \qquad (3.4.11)$$

where

$$\overline{WPK}(x,\xi) = \int_{-\infty}^{\infty} WPK(x,f) \exp\left( i2\pi f \xi \right) df \qquad (3.4.12)$$

is Fourier transform of the wave propagation kernel $WPK(.,.)$ and

$$\overline{DRK}(\xi,k) = \sum_{r=0}^{N-1} DRK(r,k)\exp\left(-i2\pi r^{(s)}\Delta f^{(s)}\xi\right) \ , \qquad (3.4.13)$$

is a Fourier series expansion with the discrete reconstruction kernel $DRK(r,k)$ as expansion coefficients.

Function $\overline{PSF}(x,\xi;k)$ does not depend on physical parameters of the hologram sampling device. It depends solely on wave propagation and discrete convolution kernels. We will refer to this function as **PSF of sampled hologram reconstruction**. Overall PSF of the numerical reconstruction of holograms $OPSF(x,k)$ and PSF of sampled hologram reconstruction $\overline{PSF}(x,\xi;k)$ are linked through the integral transform

$$OPSF(x,k) = \int_{-\infty}^{\infty} \Phi_f^{(s)}(\xi)\overline{PSF}(x,\xi;k)\,d\xi \qquad (3.4.14)$$

with frequency response of the hologram sampling device as a transform kernel.

### 3.4.2.   Point spread function of numerical reconstruction of holograms recorded in far diffraction zone (Fourier holograms)

Consider point spread function of numerical reconstruction of Fourier holograms. For far diffraction zone, wave propagation kernel is that of the integral Fourier transform:

$$WPK(x,f) = \exp\left(-i2\pi\frac{xf}{\lambda Z}\right) \ . \qquad (3.4.15)$$

Its Fourier Transform is a delta-function:

$$\overline{WPK}(x,\xi) =$$

$$\int_{-\infty}^{\infty} WPK(x,f)\exp\left(i2\pi f\xi\right)df = \int_{-\infty}^{\infty}\exp\left[-i2\pi f\left(\frac{x}{\lambda Z}-\xi\right)\right]df = \delta\left(\frac{x}{\lambda Z}-\xi\right). \ (3.4.16)$$

Assume that Shifted DFT with discrete reconstruction kernel

$$DRK(r,k) = \exp\left[i2\pi\frac{(k+u)(r+v)}{N}\right] . \qquad (3.4.16)$$

defined in Ch. 3, Sect. 3.2.2 is used for numerical reconstruction of Fourier hologram. Fourier series expansion over this discrete reconstruction kernel is:

$$\overline{DRK}(\xi,k)=\sum_{r=0}^{N-1}DRK(r,k)\exp\left(-i2\pi\tilde{r}^{(s)}\Delta f^{(s)}\xi\right)=$$

$$\sum_{r=0}^{N-1}\exp\left[i2\pi\frac{(k+u)(r+v)}{N}\right]\exp\left[-i2\pi\left(r+v^{(s)}\right)\Delta f^{(s)}\xi\right]=$$

$$\exp\left(-i2\pi v^{(s)}\Delta f^{(s)}\xi\right)\exp\left[i2\pi\frac{(k+u)v}{N}\right]\sum_{r=0}^{N-1}\exp\left[i2\pi\left(\frac{(k+u)}{N}-\Delta f^{(s)}\xi\right)r\right]=$$

$$\exp\left(-i2\pi v^{(s)}\Delta f^{(s)}\xi\right)\exp\left[i2\pi\frac{(k+u)v}{N}\right]\frac{\exp\left[i2\pi N\left(\frac{k+u}{N}-\Delta f^{(s)}\xi\right)\right]-1}{\exp\left[i2\pi\left(\frac{k+u}{N}-\Delta f^{(s)}\xi\right)\right]-1}=$$

$$\exp\left[-i2\pi\left(v^{(s)}+\frac{N-1}{2}\right)\Delta f^{(s)}\xi\right]\exp\left\{i2\pi\frac{(k+u)}{N}\left(v+\frac{N-1}{2}\right)\right\}\times$$

$$\frac{\sin\left[\pi N\left(\Delta f^{(s)}\xi-\frac{k+u}{N}\right)\right]}{\sin\left[\pi\left(\Delta f^{(s)}\xi-\frac{k+u}{N}\right)\right]}\tag{3.4.17}$$

In order to eliminate pure phase exponential multiplicands in Eq. 3.4.17, one can choose shift parameters $v^{(s)}$ and $v$ of sampling and reconstruction transform as:

$$v^{(s)}=v=-\frac{N-1}{2}.\tag{3.4.18}$$

With these shift parameters,

$$\overline{DRK}(\xi,k)=\frac{\sin\left[\pi\left(\Delta f^{(s)}\xi-\frac{(k+u)}{N}\right)N\right]}{\sin\left[\pi\left(\Delta f^{(s)}\xi-\frac{(k+u)}{N}\right)\right]}=N\,\mathrm{sincd}\left[N;\pi\left(\Delta f^{(s)}\xi-\frac{(k+u)}{N}\right)N\right]$$

$$\tag{3.4.19}$$

and

$$\overline{PSF}(x,\xi;k)=\overline{WPK}(x,\xi)\cdot\overline{DRK}(\xi,k)$$

$$N\,\mathrm{sincd}\left[N;\pi\left(\Delta f^{(s)}\xi-\frac{(k+u)}{N}\right)N\right]\delta\left(\frac{x}{\lambda Z}-\xi\right).\tag{3.4.20}$$

Then, finally, obtain that the overall point spread function of numerical reconstruction of Fourier hologram is

$$OPSF\left(x,k\right) = \int_{-\infty}^{\infty} \Phi_f^{(s)}\left(\xi\right)\overline{PSF}\left(x,\xi;k\right)d\xi =$$

$$N\,\mathrm{sincd}\left[N;\pi\left(\frac{\Delta f^{(s)}x}{\lambda Z} - \frac{(k+u)}{N}\right)N\right]\int_{-\infty}^{\infty}\Phi_f^{(s)}\left(\xi\right)\delta\left(\frac{x}{\lambda Z}-\xi\right)d\xi$$

$$= N\,\mathrm{sincd}\left[N;\pi\left(\frac{\Delta f^{(s)}x}{\lambda Z} - \frac{(k+u)}{N}\right)N\right]\Phi_f^{(s)}\left(\frac{x}{\lambda Z}\right) =$$

$$= N\,\mathrm{sincd}\left[N;\pi\left(x-(k+u)\Delta_x\right)/\Delta_x\right]\Phi_f^{(s)}\left(\frac{x}{\lambda Z}\right), \qquad (3.4.21)$$

where

$$\Delta_x = \lambda Z / N\Delta f^{(s)} = \lambda Z / S_H \qquad\qquad (3.4.22,\ a)$$

in an efficient sampling interval and

$$S_H = N\Delta f^{(s)} \qquad\qquad (3.4.22,\ b)$$

is the physical size of the hologram.

Formula (3.4.21) has a clear physical interpretation illustrated in Fig.3.3. It shows two multiplicands defining OPSF's of numerical reconstruction of holograms digitally recorded in far diffraction zone: discrete sinc-function and frequency responses of the ideal sampling device and of digital cameras for two different camera fill-factor, or the ratio of the size of camera sensitive elements to the inter-pixel distance (camera pitch).

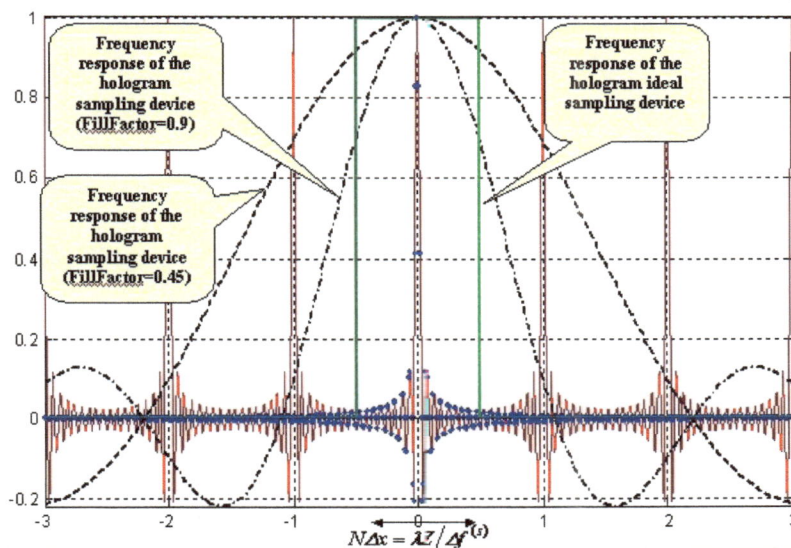

**Fig. 3.3.** Components of OPSF of numerical reconstruction of holograms digitally recorded in far diffraction zone. Thin red solid line represents a discrete sinc-function. Bold blue dots are samples of the continuous sinc-function, the ideal sampling PSF, to compare it with the discrete sinc-function. Green bold rectangle is the frequency response of the hologram ideal sampling device. Dotted lines are frequency responses of real hologram sampling devices, digital cameras with fill factors 0.9 and 0.45.

As it was mentioned, point spread function of the ideal signal sampling device is a sinc-function and its frequency response is a rect-function. Provided the hologram sampling device is such an ideal sampler with frequency response:

$$\Phi_f^{(s)}\left(\frac{x}{\lambda Z}\right) = rect\left(\frac{x + \lambda Z/2\Delta f^{(s)}}{\Delta f^{(s)}/\lambda Z}\right) = \begin{cases} 1, -\lambda Z/2\Delta f^{(s)} \le x \le \lambda Z/2\Delta f^{(s)} \\ 0, \qquad\qquad otherwise \end{cases}, \quad (3.4.23)$$

overall point spread function of numerical reconstruction of Fourier hologram is a discrete sinc-function, and object wave front samples are measured within the object's spatial extent interval $-\lambda Z/2\Delta f^{(s)} \le x \le \lambda Z/2\Delta f^{(s)}$ defined by the spread of the hologram sampling device frequency response. Therefore, with an ideal hologram sampler, numerical reconstruction of Fourier hologram is almost ideal object wave front sampling, as discrete sinc-function approximates continuous sinc-function within the interval $-\lambda Z/2\Delta f^{(s)} \le x \le \lambda Z/2\Delta f^{(s)}$ relatively closely, the closer, the larger the number of hologram samples $N$.

In reality, hologram sampling devices are, of course, not ideal samplers, and their frequency responses are not rectangular functions. They are rather not uniform within the basic object extend interval $-\lambda Z/2\Delta f^{(s)} \le x \le \lambda Z/2\Delta f^{(s)}$ and decay not abruptly outside this interval but quite gradually. As a consequence, each of the object samples is a combination of the sample, measured by the main lobe of the discrete sinc-function within the basic object extend interval and samples collected by other lobes of the discrete sinc-function outside the basic interval. This effect is called ***aliasing***. It is one source of the measurement errors. In particular, for diffusely reflecting objects, it may result in an additional speckle noise in the reconstructed object image. One can avoid this distortion if, in the process of making object hologram, object is illuminated strictly within the basic interval as defined by the hologram sampling interval (camera pitch).

The second source of the reconstruction errors is associated with non-uniformity of the hologram sampler frequency response within the basic interval. These errors can be compensated by multiplying the reconstruction results by the function inverse to the frequency response of the hologram sampler $\Phi_f^{(s)}\left(\frac{x}{\lambda Z}\right)$.

One can also see from Eqs. 3.4.21 and 3.4.22 that the resolving power of numerical reconstruction of Fourier hologram is determined by the distance $\Delta_x$ between zeros of the discrete sinc-function, which is equal to $\lambda Z/N\Delta f^{(s)} = \lambda Z/S_H$, where $S_H = N\Delta f^{(s)}$ is the physical size of the sampled hologram. Due to this finite resolving power, one can also expect, for diffuse objects, a certain amount of speckle noise in the reconstructed object.

### 3.4.3.  Point spread function of numerical reconstruction of holograms recorded in near diffraction zone (Fresnel holograms)

For near diffraction zone wave propagation kernel is, to the accuracy of irrelevant constant multiplicand,

$$WPK(x,f) = \exp\left[ i\pi \frac{(x-f)^2}{\lambda Z} \right].$$  (3.4.24)

Its Fourier Transform is:

$$\overline{WPK}(x,\xi) = \int_{-\infty}^{\infty} \exp\left[ i\pi \frac{(x-f)^2}{\lambda Z} \right] \exp(i2\pi f\xi) df =$$

$$\exp(i2\pi x\xi) \int_{-\infty}^{\infty} \exp\left( i\pi \frac{p^2}{\lambda Z} \right) \exp(-i2\pi p\xi) dp,$$  (3.4.25)

or, with an account of Eq. A.3.5.1 (Appendix A3.3, Ch. 3),

$$\overline{WPK}(x,\xi) = \exp(i2\pi x\xi)\exp(-i\pi\lambda Z\xi^2)$$  (3.4.26)

again with an irrelevant constant multiplicand omitted. In what follows, we will separately consider point spread function for Fourier and Convolution reconstruction algorithms used for reconstruction of Fresnel holograms and described in Ch. 3, Sect. 3.3. . For simplicity's sake, zero shifts will be assumed in both, hologram and object wave front domains.

### Fourier reconstruction algorithm

In the Fourier reconstruction algorithm, discrete reconstruction kernel is, with zero shifts in hologram and object wave front domains,

$$DRK(r,k) = \left[ \exp\left( -i\pi \frac{k^2\mu^2}{N} \right) \exp\left( i2\pi \frac{kr}{N} \right) \right] \exp\left( -i\pi \frac{r^2}{\mu^2 N} \right).$$  (3.4.27)

Fourier series expansion defined by Eq. 3.4.13 for this kernel is

$$\overline{DRK}(\xi,k) = \exp\left( -i\pi \frac{k^2\mu^2}{N} \right) \sum_{r=0}^{N-1} \exp\left( -i\pi \frac{r^2}{\mu^2 N} \right) \exp\left( i2\pi \frac{kr}{N} \right) \exp\left( -i2\pi r\Delta f^{(s)}\xi \right) =$$

$$\exp\left( -i\pi \frac{k^2\mu^2}{N} \right) \sum_{r=0}^{N-1} \exp\left( -i\pi \frac{r^2}{\mu^2 N} \right) \exp\left[ i2\pi r\left( \frac{k}{N} - \Delta f^{(s)}\xi \right) \right],$$  (3.4.28, a)

or, in a more compact form,

$$\overline{DRK}(\xi,k) = \exp\left( -i\pi \frac{k^2\mu^2}{N} \right) \text{frincd}\left( N; -1/\mu^2; \frac{k}{N} - \Delta f^{(s)}\xi \right).$$  (3.4.28, b)

Then obtain that PSF of sampled hologram reconstruction is

$$\overline{PSF}\left(x,\xi;k\right)=\overline{WPK}\left(x,\xi\right)\cdot\overline{DRK}\left(\xi,k\right)\propto$$

$$\exp\left(i2\pi x\xi\right)\exp\left(-i\pi\lambda Z\xi^2\right)\exp\left(-i\pi\frac{k^2\mu^2}{N}\right)\mathrm{frincd}\left(N;-1/\mu^2;\frac{k}{N}-\Delta f^{(s)}\xi\right) \quad (3.4.29)$$

and that OPSF of numerical reconstruction of Fresnel holograms with Fourier reconstruction algorithm is:

$$OPSF\left(x,k\right)=\exp\left(-i\pi\frac{k^2\mu^2}{N}\right)\times$$

$$\int_{-\infty}^{\infty}\Phi_f^{(s)}\left(\xi\right)\exp\left(-i\pi\lambda Z\xi^2\right)\mathrm{frincd}^*\left(N;1/\mu^2;\frac{k}{N}-\Delta f^{(s)}\xi\right)\exp\left(i2\pi x\xi\right)d\xi \quad (3.4.30)$$

Eqs. 3.4.30 is much more involved for an analytical treatment that the corresponding equation for OPSF of numerical reconstruction of Fourier holograms and, in general, requires numerical methods for analysis. In order to facilitate its treatment, rewrite Eq. 3.4.30 using Eq. 3.4.28, a) for $\overline{DRK}\left(\xi,k\right)$:

$$OPSF\left(x,k\right)=\exp\left[i\pi\left(\frac{x^2}{\lambda Z}-\frac{k^2\mu^2}{N}\right)\right]\times$$

$$\int_{-\infty}^{\infty}\Phi_f^{(s)}\left(\xi\right)\sum_{r=0}^{N-1}\exp\left(-i\pi\frac{r^2}{\mu^2 N}\right)\exp\left[i2\pi r\left(\frac{k}{N}-\Delta f^{(s)}\xi\right)\right]\exp\left[-i\pi\frac{\left(x-\lambda Z\xi\right)^2}{\lambda Z}\right]d\xi=$$

$$\exp\left[i\pi\left(\frac{x^2}{\lambda Z}-\frac{k^2\mu^2}{N}\right)\right]\sum_{r=0}^{N-1}\exp\left(-i\pi\frac{r^2}{\mu^2 N}\right)\exp\left(i2\pi\frac{kr}{N}\right)\times$$

$$\int_{-\infty}^{\infty}\Phi_f^{(s)}\left(\xi\right)\exp\left[-i\pi\frac{\left(x-\lambda Z\xi\right)^2}{\lambda Z}\right]\exp\left(-i2\pi r\Delta f^{(s)}\xi\right)d\xi \quad (3.4.31)$$

Assume now that frequency response of the hologram sampling device $\Phi_f^{(s)}\left(\xi\right)$ is constant, which is equivalent to the assumption that its PSF is a delta function. Practically, this means that the hologram recording photographic camera is assumed to have a very small fill-factor.   In this simplifying assumption,

$$OPSF\left(x,k\right)=\exp\left[i\pi\left(\frac{x^2}{\lambda Z}-\frac{k^2\mu^2}{N}\right)\right]\times$$

$$\sum_{r=0}^{N-1}\exp\left(-i\pi\frac{r^2}{\mu^2 N}\right)\exp\left(i2\pi\frac{kr}{N}\right)\int_{-\infty}^{\infty}\exp\left[-i\pi\frac{\left(x-\lambda Z\xi\right)^2}{\lambda Z}\right]\exp\left(-i2\pi r\Delta f^{(s)}\xi\right)d\xi=$$

$$\exp\left[i\pi\left(\frac{x^2}{\lambda Z}-\frac{k^2\mu^2}{N}\right)\right]\times$$

$$\sum_{r=0}^{N-1}\exp\left(-i\pi\frac{r^2}{\mu^2 N}\right)\exp\left(i2\pi\frac{kr}{N}\right)\int_{-\infty}^{\infty}\exp\left[-i\pi\frac{(x-\lambda Z\xi)^2}{\lambda Z}\right]\exp\left(-i2\pi r\Delta f^{(s)}\xi\right)d\xi=$$

$$\exp\left[i\pi\left(\frac{x^2}{\lambda Z}-\frac{k^2\mu^2}{N}\right)\right]\times$$

$$\sum_{r=0}^{N-1}\exp\left(-i\pi\frac{r^2}{\mu^2 N}\right)\exp\left(i2\pi\frac{kr}{N}\right)\int_{-\infty}^{\infty}\exp\left(-i\pi\frac{\tilde{\xi}^2}{\lambda Z}\right)\exp\left(-i2\pi r\Delta f^{(s)}\frac{x-\tilde{\xi}}{\lambda Z}\right)d\xi=$$

$$\exp\left[i\pi\left(\frac{x^2}{\lambda Z}-\frac{k^2\mu^2}{N}\right)\right]\times\sum_{r=0}^{N-1}\exp\left(-i\pi\frac{r^2}{\mu^2 N}\right)\exp\left(i2\pi\frac{kr}{N}\right)\exp\left(-i2\pi r\Delta f^{(s)}\frac{x}{\lambda Z}\right)\times$$

$$\int_{-\infty}^{\infty}\exp\left(-i\pi\frac{\tilde{\xi}^2}{\lambda Z}\right)\exp\left(i2\pi\frac{r\Delta f^{(s)}}{\lambda Z}\tilde{\xi}\right)d\xi\propto\exp\left[i\pi\left(\frac{x^2}{\lambda Z}-\frac{k^2\mu^2}{N}\right)\right]\times$$

$$\sum_{r=0}^{N-1}\exp\left(-i\pi\frac{r^2}{\mu^2 N}\right)\exp\left[i2\pi\left(\frac{k}{N}-\frac{\Delta f^{(s)}x}{\lambda Z}\right)r\right]\exp\left(i\pi\frac{r^2\Delta f^{(s)2}}{\lambda Z}\right)=$$

$$\exp\left[i\pi\left(\frac{x^2}{\lambda Z}-\frac{k^2\mu^2}{N}\right)\right]\sum_{r=0}^{N-1}\exp\left[-i\pi\left(\frac{N\Delta f^{(s)2}}{\lambda Z}-\frac{1}{\mu^2}\right)\frac{r^2}{N}\right]\exp\left[i2\pi\left(\frac{k}{N}-\frac{\Delta f^{(s)}x}{\lambda Z}\right)r\right],$$

(3.4.32 a)

or

$$OPSF(x,k)=\exp\left[i\pi\left(\frac{x^2}{\lambda Z}-\frac{k^2\mu^2}{N}\right)\right]\mathrm{frincd}^*\left(N;\frac{N\Delta f^{(s)2}}{\lambda Z}-\frac{1}{\mu^2};\frac{k}{N}-\frac{\Delta f^{(s)}x}{\lambda Z}\right).$$

(3.4.32, b)

As one can see from Eq. 3.4.32, b), OPSF of numerical reconstruction of Fresnel holograms recorded with cameras with very small fill-factor is just proportional to frincd-function defined in Ch.2 (Eqs. 2.3.12 and Figs. 2.4 and 2.5)

An important special case is "in focus" reconstruction, when

$$\mu^2=\frac{\lambda Z}{N\Delta f^{(s)2}}$$

(3.4.33)

In this case numerical reconstruction point spread function is discrete sinc-function

$$OPSF(x,k)=\exp\left[i\frac{\lambda Z}{\Delta f^{(s)2}}\left(\frac{\Delta f^{(s)2}x^2}{\lambda^2 Z^2}-\frac{k^2}{N^2}\right)\right]\mathrm{sincd}\left[N;\pi\left(\frac{\Delta f^{(s)}x}{\lambda Z}-\frac{k}{N}\right)N\right]=$$

$$\exp\left[i\frac{\lambda Z}{\Delta f^{(s)2}}\left(\frac{\Delta f^{(s)2}x^2}{\lambda^2 Z^2}-\frac{k^2}{N^2}\right)\right]\mathrm{sincd}\left[N;\pi(x-k\Delta_x)/\Delta_x\right]$$

(3.4.34)

where $\Delta_x = \lambda Z / N \Delta f^{(s)} = \lambda Z / S_H$. As one can see, "in focus" reconstruction PSF is essentially the same as that of numerical reconstruction of Fourier holograms (Eq. (3.4.22) for the same assumption regarding the hologram sampling device. It has the same resolving power and provides aliasing free object reconstruction within the interval $S_o = \lambda Z / \Delta f^{(s)}$.

One can establish a link between the size of this interval and the value $\mu^2 = \lambda Z / N \Delta f^{(s)2} = \lambda Z N / S_H^2$ of the focusing parameter required for the reconstruction:

$$ S_o = \lambda Z / \Delta f^{(s)} = \lambda Z N / S_H = \mu^2 S_H \qquad (3.4.35) $$

From this relationship it follows that aliasing free reconstruction of the object from a hologram recorded on a distance defined by the focusing parameter $\mu^2$ is possible if the object size does not exceed the value $\mu^2 S_H$. Therefore, for $\mu^2 < 1$, allowed object size should be less then the hologram size otherwise aliasing caused by the periodicity of the discrete sinc-function will appear.

**Convolution reconstruction algorithm**

In the convolution reconstruction algorithm, discrete reconstruction kernel is, with zero shifts in hologram and object wave front domains,

$$ DRK(r,k) = \mathrm{frincd}\left(N;\mu^2;k-r\right) = \frac{1}{N}\sum_{s=0}^{N-1}\exp\left(i\pi\frac{\mu^2 s^2}{N}\right)\exp\left[-i2\pi\frac{(k-r)s}{N}\right] (3.4.36) $$

Fourier series expansion defined by Eq. 3.4.13 for this kernel is

$$ \overline{DRK}(\xi,k) = \frac{1}{N}\sum_{r=0}^{N-1}\sum_{s=0}^{N-1}\exp\left(i\pi\frac{\mu^2 s^2}{N}\right)\exp\left[-i2\pi\frac{(k-r)s}{N}\right]\exp\left(-i2\pi r\Delta f^{(s)}\xi\right) = $$

$$ \frac{1}{N}\sum_{s=0}^{N-1}\exp\left(i\pi\frac{\mu^2 s^2}{N}\right)\exp\left(-i2\pi\frac{ks}{N}\right)\sum_{r=0}^{N-1}\exp\left[i2\pi r\left(\frac{s}{N}-\Delta f^{(s)}\xi\right)\right]. \qquad (3.4.37) $$

Then obtain that PSF of sampled hologram reconstruction is

$$ \overline{PSF}(x,\xi;k) = \overline{WPK}(x,\xi)\cdot\overline{DRK}(\xi,k) \propto \frac{\exp(i2\pi x\xi)\exp(-i\pi\lambda Z\xi^2)}{N} \times $$

$$ \sum_{s=0}^{N-1}\exp\left(i\pi\frac{\mu^2 s^2}{N}\right)\exp\left(-i2\pi\frac{ks}{N}\right)\sum_{r=0}^{N-1}\exp\left[i2\pi r\left(\frac{s}{N}-\Delta f^{(s)}\xi\right)\right] \qquad (3.4.38) $$

and that PSF of numerical reconstruction of Fresnel holograms with Fourier reconstruction algorithm is:

$$PSF(x,k) = \int_{-\infty}^{\infty} \Phi_f^{(s)}(\xi) \exp(-i\pi\lambda Z \xi^2) \exp(i2\pi x \xi) d\xi \times$$

$$\frac{1}{N} \sum_{s=0}^{N-1} \exp\left(i\pi \frac{\mu^2 s^2}{N}\right) \exp\left(-i2\pi \frac{ks}{N}\right) \sum_{r=0}^{N-1} \exp\left[i2\pi r\left(\frac{s}{N} - \Delta f^{(s)}\xi\right)\right] =$$

$$\int_{-\infty}^{\infty} \Phi_f^{(s)}(\xi) \exp(-i\pi\lambda Z \xi^2) \exp(i2\pi x \xi) d\xi \sum_{s=0}^{N-1} \exp\left(i\pi \frac{\mu^2 s^2}{N}\right) \exp\left(-i2\pi \frac{ks}{N}\right) \times$$

$$\frac{\sin\left[\pi\left(s - N\Delta f^{(s)}\xi\right)\right]}{N\sin\left[\dfrac{\pi}{N}\left(s - N\Delta f^{(s)}\xi\right)\right]} \exp\left[i\pi \frac{N-1}{N}\left(s - N\Delta f^{(s)}\xi\right)\right], \qquad (3.4.39)$$

or finally,

$$PSF(x,k) =$$

$$\int_{-\infty}^{\infty} \Phi_f^{(s)}(\xi) \exp(-i\pi\lambda Z \xi^2) \exp(i2\pi x \xi) d\xi \sum_{s=0}^{N-1} \exp\left(i\pi \frac{\mu^2 s^2}{N}\right) \exp\left(-i2\pi \frac{ks}{N}\right) \times$$

$$\mathrm{sincd}\left[N; \pi\left(s - N\Delta f^{(s)}\xi\right)\right]. \qquad (3.4.40)$$

Note that the last phase exponential multiplicand in Eq. 3.4.40 is omitted, as it appears due to the assumption of zero shift parameters in the reconstruction algorithm.

Although Eq. 3.4.40 is quite nontransparent for analysis, at least one important property can be immediately seen from it, that of periodicity of the PSF over object sample index $k$ with period $N$. As, by the definition of the Convolutional Fresnel Transform, sampling interval $\Delta x^{(r)}$ in the object plane is identical to the hologram sampling interval $\Delta f^{(s)}$, this periodicity of the PSF implies that object wave front is reconstructed within the physical interval $N\Delta x = N\Delta f^{(s)} = S_H$, where $S_H$ is the physical size of the hologram. Further detailed analysis is not feasible without bringing in numerical methods. Some results of such a numerical analysis[2] are illustrated in Fig. 3.4.

Figs. 3.4, a) and b) reveal, in particular, that, although object sampling interval in the convolution method is deliberately set equal to the hologram sampling interval, resolving power of the method is still defined by the same fundamental value $\lambda Z/S_H$ as that of the Fourier reconstruction algorithm and of the Fourier reconstruction algorithm for Fourier holograms. One can clearly see this when one compares width of the main lobe of the point spread function in Fig. 3.4, a) with the distance between vertical ticks that indicate object sampling positions and from observing, in Fig. 3.4, b), three times widening of the width of the main lobe of PSF that corresponds to the object-to-hologram distance parameter $\mu^2 = 0.45$ with respect to that for $\mu^2 = 0.15$. Fig. 3.4, c) shows reconstruction of nine point sources placed uniformly within the object size. The plot vividly demonstrates that the hologram sampling device point

---

[2] The analysis was carried out with the help of Dr. Fucai Zhang, Institute of Technical Optics, Stuttgart University, Germany

spread function acts very similarly to its action in the case of the Fourier reconstruction algorithm and of reconstruction of Fourier holograms: it modulates the reconstruction result with a function close to its Fourier transform, the frequency response of the hologram sampling device.

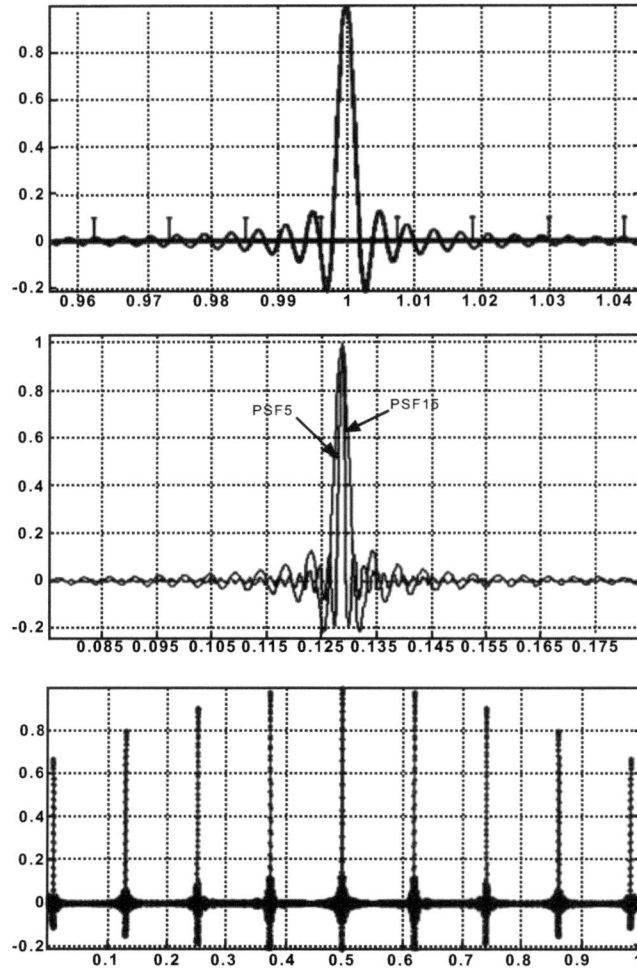

**Fig. 3.4.** To the explanation of point spread functions of the convolution algorithm: central lobe of OPSF shown along with boundaries of object sampling interval shown by vertical ticks (top); central lobes of OPSF of reconstructions for two distances between object and hologram (*PSF5* and *PSF*15, middle); c) reconstruction result for 9 point sources placed uniformly within object area (bottom).

Finite width of the reconstruction algorithm PSFs does not only limits their resolving power. It also causes speckle noise in the reconstruction of diffuse objects for the same reason that was mentioned above in the discussion of OPSF of reconstruction of Fourier holograms. This phenomenon is illustrated in Fig. 3.5 that compares reconstructions of intensity of rectangular shaped non-diffuse and diffuse objects.

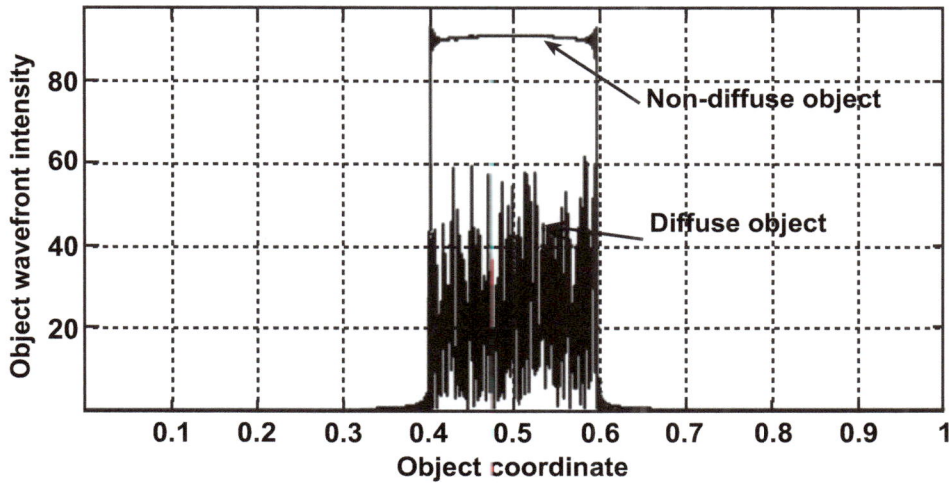

**Fig. 3.5.** Results of computer simulation reconstruction by the convolution algorithm of a Fresnel hologram of non-diffuse and diffuse objects that demonstrate appearance of heavy speckle noise for the diffuse object

## 3.5 Zones of applicability of the Fourier and convolution reconstruction algorithms

From the above analysis of point spread functions of the Fourier and Convolution algorithms, as well as from properties of the frincd-function introduced in Ch. 3, Sect. 3.3.5 it follows that the algorithms has different zones of application in terms of the distance parameter $\mu^2 = \lambda Z / N \Delta f^{(s)2} = \lambda Z N / S_H^2$ that connects wave length of the object illumination $\lambda$, distance between the object and hologram planes $Z$, the number of hologram samples $N$, pitch of the hologram recording camera $\Delta f^{(s)2}$ and its size $S_H = \Delta f^{(s)}$ (for the definition of frincd-function given by Eq. 3.3.12, the corresponding parameter is $q = 1/\mu^2$).

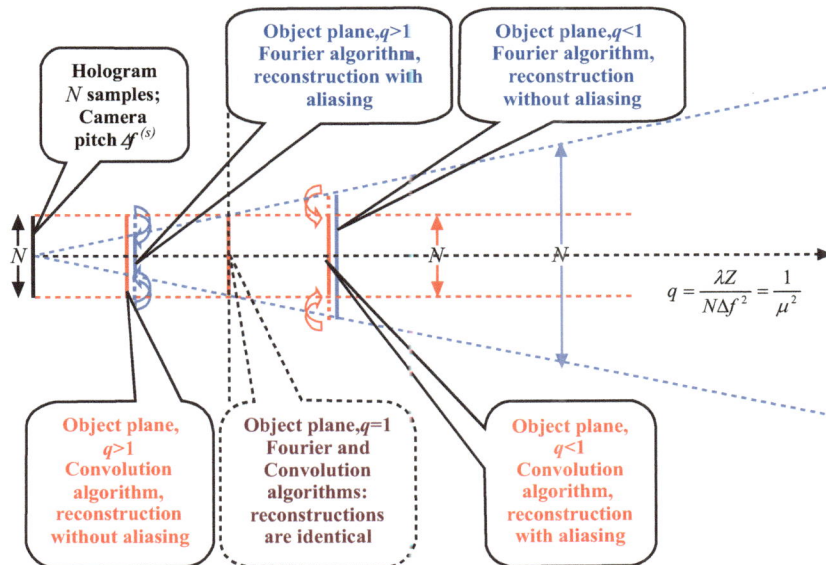

**Fig. 3.6.** Zones of applicability of Fourier and Convolution algorithms for numerical reconstruction of Fresnel holograms. Round arrows indicate appearance of aliasing.

Specifically, Fourier reconstruction algorithm provides aliasing-free reconstruction for distances that are large enough to secure that $\mu^2 \geq 1$ ($q \leq 1$) while the convolution algorithm works aliasing-free reconstruction for closer distances when $\mu^2 \leq 1$ ($q \geq 1$). When $\mu^2 = 1$ ($q = 1$) both methods give identical results. These two zones are sketched in Fig. 3.6.

The aliasing phenomena in numerical reconstruction of digitally recorded holograms are illustrated in a movie (Fig. 3.7, a) and in Fig. 3.8. The movie illustrates results of reconstruction of a Fresnel hologram of a test object sketched in Fig. 3.7, b), which consists of two crossed thin wires and a ruler installed at different distances from the hologram[3]. Fig. 3.8 shows several particular frames of this movie obtained for focusing parameter $\mu^2 = 0.2439$, $\mu^2 = 0.6618$ and $\mu^2 = 1$. One can clearly see in Fig. 3.8, aliasing artifacts in the Fourier reconstruction due to overlapping of reconstructions from two side lobes of the discrete sinc-functions for $\mu^2 = 0.2439$, $\mu^2 = 0.6618$. These artifacts disappear if outer parts of the hologram outside the circle with diameter equal to $\mu^2$-th fraction of the hologram size are set to zeros before applying the reconstruction algorithm. This can be seen form the result of reconstruction of such a trimmed hologram shown in Fig. 3.8, middle column. Note that the Fourier reconstruction algorithm with above mentioned zeroing of the excessive part of the hologram for $\mu^2 < 1$ acts, with respect to the convolution algorithm reconstruction, as a "magnifying glass". Object aliasing artifacts of the Fourier reconstruction algorithm for $\mu^2 < 1$ can also be avoided if the hologram is reconstructed by fragments of $\mu^2 S_H$ size.

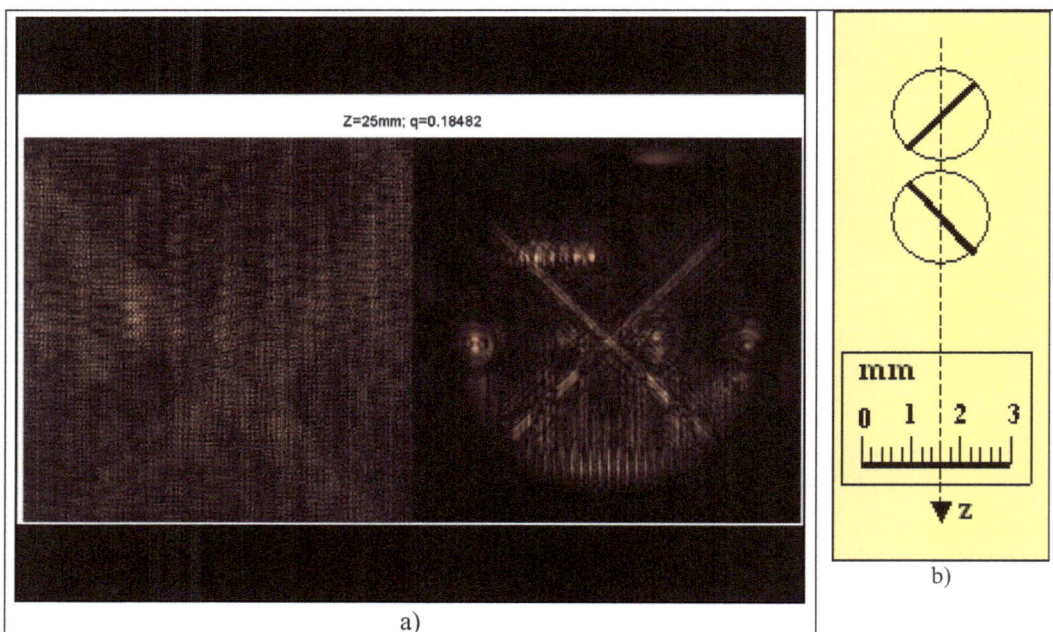

**Fig. 3.7.** A movie (a) illustrating reconstruction of a Fresnel hologram of a test object (b) by means of Fourier (left) and Convolution (right) algorithms.

[3] The hologram courtesy to Dr. J. Campos, Autonomous University of Barcelona, Bellaterra, Barcelona, Spain

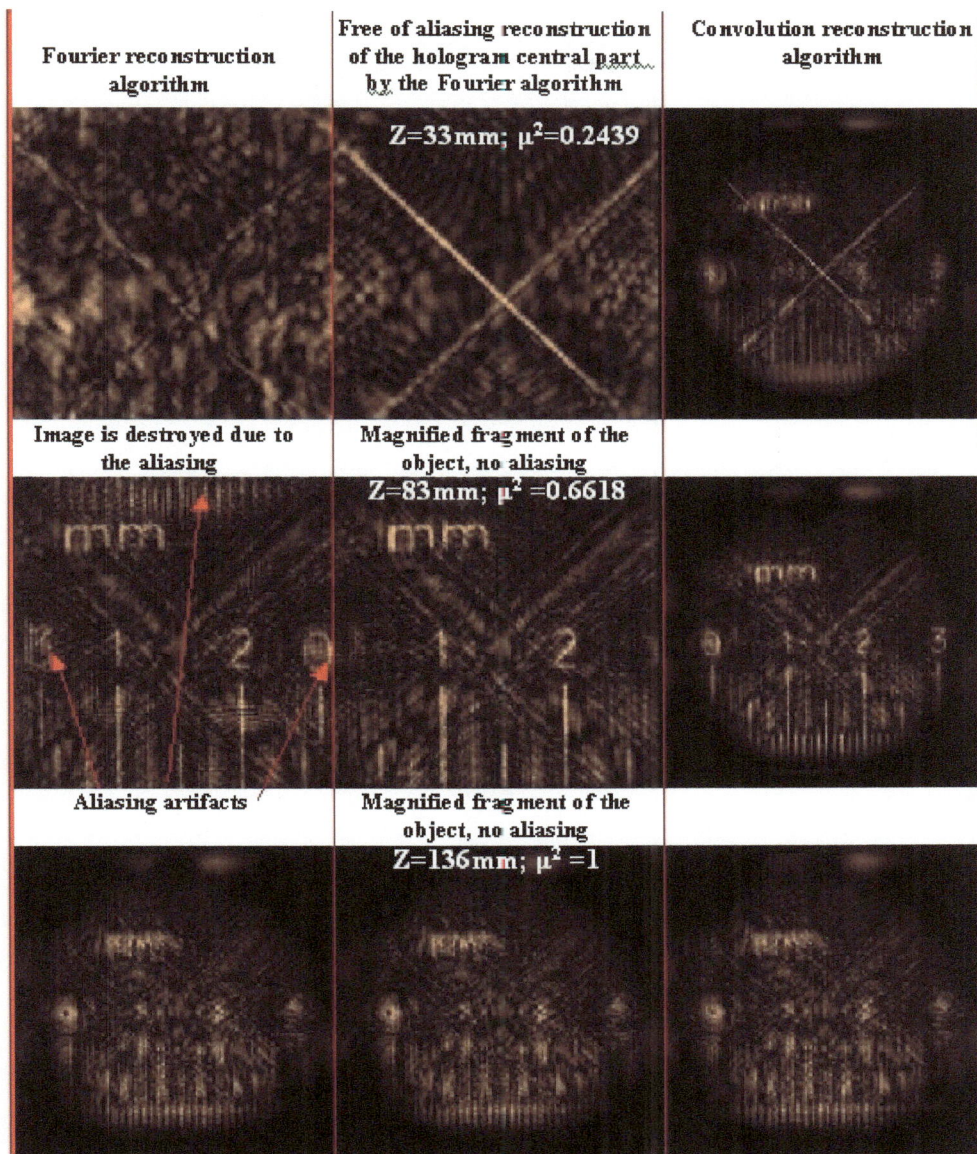

| Fourier reconstruction algorithm | Free of aliasing reconstruction of the hologram central part by the Fourier algorithm | Convolution reconstruction algorithm |
|---|---|---|
| | $Z=33mm; \mu^2=0.2439$ | |
| Image is destroyed due to the aliasing | Magnified fragment of the object, no aliasing | |
| | $Z=83mm; \mu^2=0.6618$ | |
| Aliasing artifacts | Magnified fragment of the object, no aliasing | |
| | $Z=136mm; \mu^2=1$ | |
| | All three reconstructions are identical | |

**Fig. 3.8.** Three particular results of reconstruction of the Fresnel hologram of the test object shown in Fig. 3.7 for different values of the distance parameter $\mu^2$.

# References

1.  J. W. Goodman and R. W. Lawrence, Digital image formation from electronically detected holograms, Applied Physics Letters, August 1, 1967, Volume 11, Issue 3, pp. 77-79
2.  M.A. Kronrod, N.S. Merzlyakov, L.P. Yaroslavsky, Reconstruction of a Hologram with a Computer, Soviet Physics-Technical Physics, v. 17, no. 2, 1972, p. 419 - 420.
3.  L. P. Yaroslavskii, N.S. Merzlyakov, Methods of Digital Holography, Consultance Bureau, N.Y., 1980 (English translation from Russian edition: L. P. Yaroslavskii, N.S. Merzlyakov, Methods of Digital Holography, Moscow, Izdatel'stvo Nauka, 1977. 192 p.)
4.  U. Schnars and W. Jüptner, Direct recording of holograms by a CCD target and numerical reconstruction, Applied Optics, Vol. 33, No. 2,10 January 1994, pp. 179-181

5.  E. N. Leith, J. Upatnieks, New techniques in wave front reconstruction, *JOSA*, 51, 1469-1473, 1961

6.  I. Yamaguchi, T. Zhang, Phase-shifting digital holography, Optics Letters, Vol. 22, No. 16 / August 15, 1997, 1268-1270

7.  L. Yaroslavsky, Digital Holography and Digital Image Processing, Kluwer Academic Publishers, Boston, 2004

# 4.    Computer Generated Holograms (CGH): Principles

**Abstract:** This chapter introduces basic principles and mathematical models of computer-generated holography and reviews spatial light modulators for recording of computer-generated holograms

## 4.1    Introduction

The origins of digital holography date back to 1960-70th ([1-5]). First computer generated holograms were ***binary holograms*** invented by A. Lohmann ([1-2]). They were printed on a computer line printer, then they were optically reduced and reconstructed in optical set ups using coherent laser illumination. Fig. 4.1 shows one of the first computer-generated hologram and its reconstruction. Then the use of more sophisticated devices capable of recording computer generated grey scale images was suggested for recording computer generated holograms and computer generated holograms were produced that reconstructed grey scale images of good quality ([3,4]). These grey scale holograms were recorded using an electro-mechanical rotated drum recording device (a modified faximile machine) with resolution of 200 μm ([6]) and then optically reduced with reduction factor 20 to final resolution of 10 μm for optical reconstruction with a laser beam. Fig. 4.2. represents an example of such a grey-scale computer-generated hologram and a reconstructed image.

**Fig. 4.1.** Byron Brown completing one of the first computer-generated holograms (left image) and an early CGH and its reconstruction (right image) (adopted from Ref. [5])

**Fig. 4.2.** One of the first gray scale computer-generated hologram (left) and an image reconstructed from a gray scale computer-generated hologram (right, adopted from [4])

This chapter outlines principles of computer-generated holography. In Sect. 4.2, mathematical models for synthesis of computer-generated holograms are formulated for computer-generated display holograms. In Sect. 4.3, the problem of recording computers-generated holograms (CGH) on physical media is discussed and a classification of spatial light modulators for recording CGHs is provided. Methods of encoding of computer-generated hologram for recording on physical media are given in Ch. 5.

## 4.2 Mathematical models

Basic stages in the synthesis of computer-generated holograms are outlined in a flow diagram in Fig. 4.3. These are:

- (i) Formulating mathematical models of the object and of the usage of the hologram;

- (ii) Computing the ***mathematical hologram***, array of complex numbers that represent amplitudes and phases of hologram samples in the hologram plane;

- (iii) Encoding samples of the mathematical hologram for recording them on the physical medium. At this stage, which we refer to as ***hologram encoding***, mathematical holograms are converted into arrays of numbers that control optical properties of the physical recording medium used for hologram recording.

- (iv) Fabrication of the computer generated hologram.

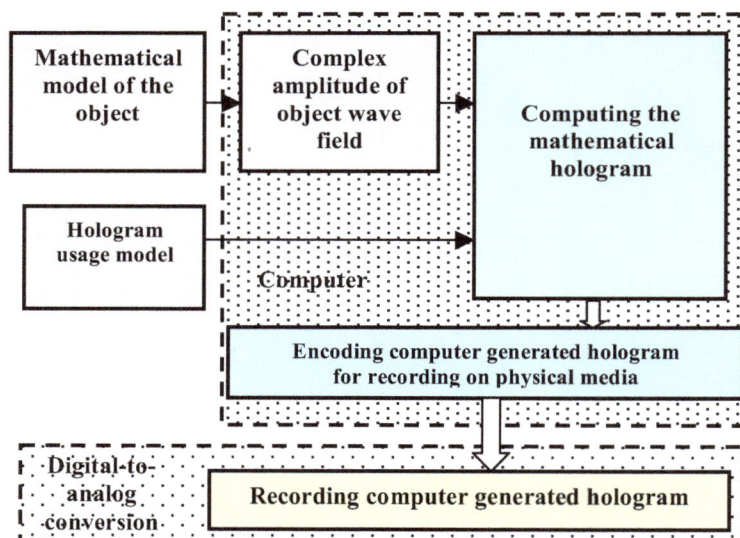

**Fig. 4.3.** Basic stages in synthesis of computer generated holograms

Computer-generated holograms can be used as spatial filters for optical information processing (to be discussed in Ch. 8), as computer-generated diffractive optical elements, such as beam forming elements (for instance, for laser tweezers), laser focusers, deflectors, beam splitters and multiplicators, and for information display (to be discussed in Ch. 9).

Mathematical models of the usage of holograms to be generated describe the geometry of wave propagation from objects to the hologram plane and specify criteria for evaluating hologram performance.

Mathematical model of the object is intended to specify the amplitude and phase distribution of the object wave front. Three types of the models can be distinguished:
- Analytical models specified by mathematical formulas or computational algorithms.
- "Geometrical", or "vector graphics" models that represent objects as compositions of elementary diffracting elements such as point scatters, segments of 2-D or 3-D curves, slits, etc; examples of such "vector graphics" models are, for instance, polygonal mesh or wire-frame models in 3D computer graphics.
- "Raster graphics", or "bitmap" models that represent objects as 2-D or 3-D arrays of object samples (pixels or voxels).

Analytical models are used mainly for generating diffractive optical elements. With analytically specified desired distribution of the object wave front, synthetic diffractive optical element is computed using methods of analytical or numerical integration with appropriate oversampling to avoid sampling artifacts of numerical integration and then sampled to generate their sampled version for recording on physical media. Methods for generating diffractive optical elements using analytical models are discussed elsewhere (see, for instance, [7]).

When objects are represented using "geometrical" or "vector graphics" models, as in Fig. 4.4, a), mathematical holograms are computed as a superposition of elemental holograms that correspond to the elements of the model (point scatters, edges of wire frame models, faces of polygonal mesh models, etc.). These elemental holograms can frequently be pre-computed and stored in a look up tables. This allows substantial reducing the computational complexity of calculation of mathematical holograms.

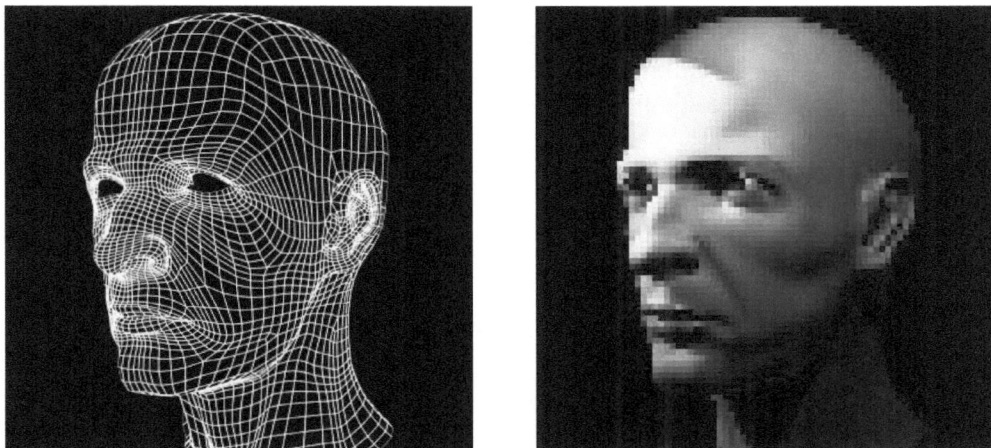

**Fig. 4.4.** Human head: "vector graphic" model (left) and "raster graphic" model (right)

"Raster graphic" models exemplified in Fig. 4.4, b) are the most general. They are better suited to object data provided by physical sensors, such as image and 3D scanners. In what follows, we will describe mathematical models for computer generated display holograms of objects specified by "raster graphic" models.

Consider a schematic diagram of visual observation of objects shown in Fig.4.5. The observer's position with respect to the observed object is defined by the observation surface where the observer's eyes are situated. The set of observation positions is defined by the object observation angle. In order to allow the observer to see the object at the given observation angle, it suffices to reproduce, by means of a hologram, the distribution of intensities and phases of the light waves scattered by the object to the observation surface.

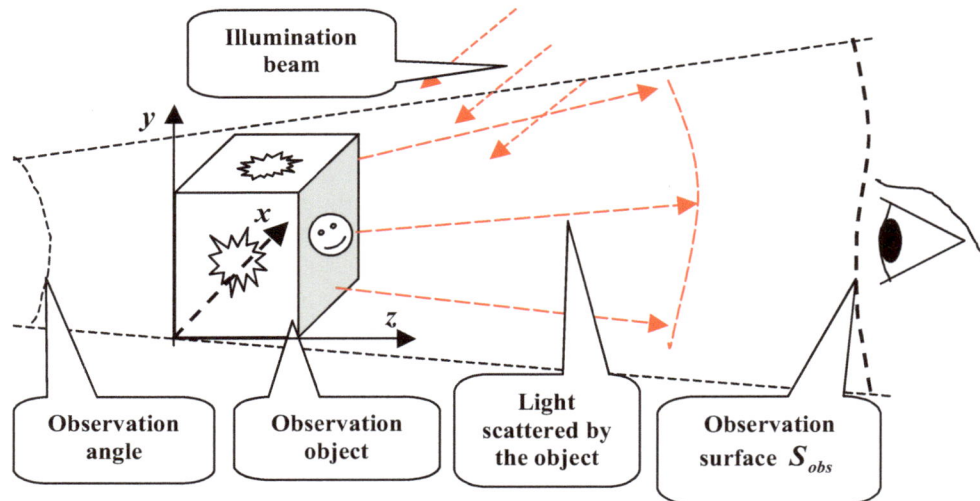

**Fig. 4.5**. Scheme of visual observation of objects by means of computer generated holograms

We will assume monochromatic illumination of objects, which enables one to describe light wave transformations in terms of the wave amplitude and phase. The case of polychromatic illumination may be treated as superposition of monochromatic illumination with different wavelengths. For instance, generating color holograms assumes generating three color separated (red-green-blue) holograms.

In monochromatic illumination, object characteristics that define its ability to reflect, transmit or scatter incident radiation may be described for our purposes by a radiation reflection or transmission factor with respect to the light complex amplitude $o(x,y,z)$ or intensity $O(x,y,z) = |o(x,y,z)|^2$ as functions of the object's surface coordinates $(x,y,z)$. Then the intensity of the scattered wave $I_{Obj}(x,y,z)$ and its complex amplitude $A_{Obj}(x,y,z)$ at the point $(x,y,z)$ of the object are related to the intensity $I(x,y,z)$ and complex amplitude $A(x,y,z)$ of the incident wave as follows:

$$I_{Obj}(x,y,z) = O(x,y,z)I(x,y,z)$$
$$A_{Obj}(x,y,z) = o(x,y,z)A(x,y,z)$$

(4.2.1)

The amplitude reflection (transmission) factor $o(x,y,z)$ may be regarded as a complex function of the spatial coordinates represented as

$$o(x,y,z) = |o(x,y,z)|\exp\left[i\theta_{Obj}(x,y,z)\right].$$

(4.2.2)

Its modulus $|o(x,y,z)|$ and phase $\theta_{Obj}(x,y,z)$ show how the modulus $|A(x,y,z)|$ and the phase $\omega(x,y,z)$ of the complex amplitude $|A(x,y,z)|\exp\left[i\omega(x,y,z)\right]$ of incident light are changed after reflection by the object surface or transmission through the object at the point $(x,y,z)$:

$$|A_{obj}(x,y,z)| = |A(x,y,z)||o(x,y,z)|$$
$$\omega_{obj}(x,y,z) = \omega(x,y,z) + \theta_{obj}(x,y,z)$$

(4.2.3)

where $|A_{obj}(x,y,z)|$ and $\omega_{obj}(x,y,z)$ are, respectively, the modulus and the phase of the object wave front.

The relation between the complex amplitude $\Gamma(\xi,\eta,\zeta)$ of the light wave field over an arbitrary surface defined by its coordinates $(\xi,\eta,\zeta)$ and the complex amplitude $A_{Obj}(x,y,z)$ of the wave front at the object surface can by described by a wave propagation integral over the object surface or volume $S_{obj}$

$$\Gamma(\xi,\eta,\zeta) = \iiint_{S_{obj}} A_{obj}(x,y,z)T(x,y,z;\xi,\eta,\zeta)\,dx\,dy\,dz \ ,$$

(4.2.4)

The kernel $T(x,y,z;\xi,\eta,\zeta)$ of the wave propagation integral depends on the spatial disposition of the object and observation surface. Reconstruction of the object can be described by a back propagation integral over the observation surface $S_{obs}$:

$$\tilde{A}_{obj}(x,y,z) = \iiint_{S_{obs}} \Gamma(\xi,\eta,\zeta)\tilde{T}(\xi,\eta,\zeta;x,y,z)\,d\xi\,d\eta\,d\zeta,$$

(4.2.5)

where $\tilde{A}_{obj}(x,y,z)$ is complex amplitude of the object as it is seen from the observation surface and $\tilde{T}(\xi,\eta,\zeta;x,y,z)$ is a kernel reciprocal to $T(x,y,z;\xi,\eta,\zeta)$.

Thus, hologram synthesis requires computation of the amplitude and phase distribution of the hologram $\Gamma(\xi,\eta,\zeta)$ for a given object wave front $A_{obj}(x,y,z)$, which is to be defined by the object description and by illumination conditions. Results of computation should then be recorded on a physical medium in a form that enables its interaction with illumination radiation for reconstructing $\tilde{A}_o\left(x,y,z\right)$ according to Eq. (4.2.5) in a form suitable for visual observation.

3-D integration required by Eq. 4.2.4 can be reduced to much more simple 2-D integration by taking into consideration the following natural limitations of visual observation:

- The pupil of the observer's eye is usually much smaller than the distance to the observation surface.
- The area of the observation surface is large compared to the inter-pupil distance of two eyes.
- The depth of objects situated at a distance comfortable for visual observation is usually small compared to the distance to the observation surface.

In view of these limitations, one can consider the observation surface as consisted of relatively small sub-areas approximated by planes. Additionally one can, using the laws of geometrical optics, replace the wave amplitude and phase distributions over the object surface by those of the wave on a plane tangent to the object (or sufficiently close to it, so as to make diffraction effects negligible) and parallel to the given plane sub-area of the observation surface. These approximations are illustrated in Fig. 4.6.

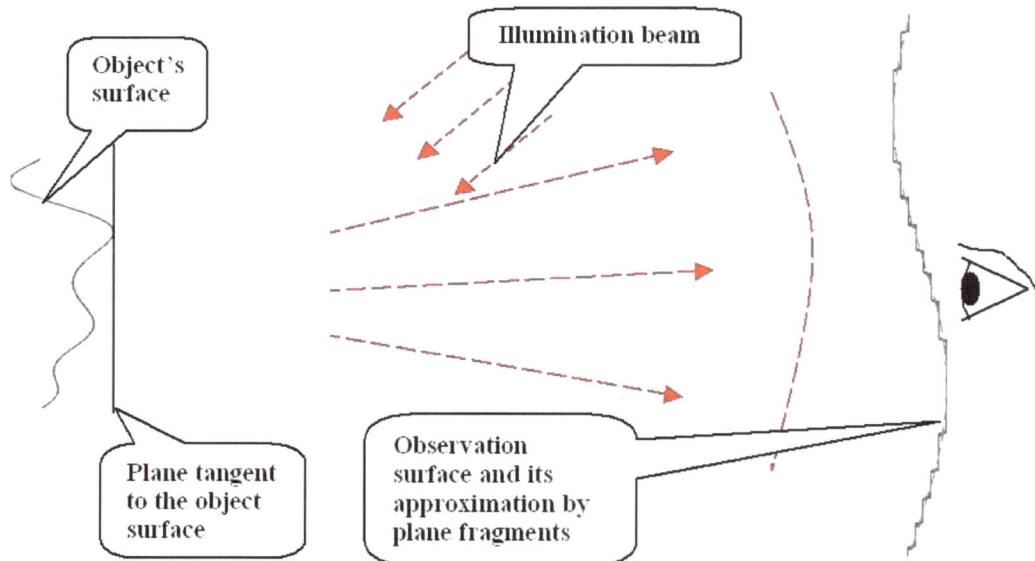

**Fig. 4.6.** Approximations of the scene and observation geometry that lead to Eq. 4.2.6.

If the geometrical dimensions of the observation object and of the observation surface fragment are small compared to the distance $Z$ from the object to the observation plane, the integral relationship of Eq. 4.2.6 is, as it was shown in Ch. 2, Sec.2.2 , reduced to the integral Fresnel transform:

$$\overline{\Gamma}_{Fr}\left(\xi,\eta\right)=\int\limits_{-\infty}^{\infty}\int\limits_{-\infty}^{\infty}\overline{A}_{o}\left(x,y\right)\exp\left\{i\pi\left[\left(x-\xi\right)^{2}+\left(y-\eta\right)^{2}\right]\Big/\lambda Z\right\}dx\,dy\,,\qquad(4.2.7)$$

where $\lambda$ is the illumination wave length. Discrete representations of the integral Fresnel transform suitable for its numerical computing is discussed in Ch. 3, Sect. 3.3. Holograms synthesized using this relationship will be referred to as synthesized ***Fresnel holograms***.

Further simplification is possible if the X-Y size of the object is small enough to satisfy the relationship

$$\pi\left(x^{2}+y^{2}\right)\Big/\lambda Z<<1,\qquad(4.2.8)$$

In this case the integral of Eq.(4.2.7) can be approximated as

$$\overline{\Gamma}_{Fr}\left(\xi,\eta\right)\cong\exp\left[i\pi\left(\xi^{2}+\eta^{2}\right)\right]\int\limits_{-\infty}^{\infty}\int\limits_{-\infty}^{\infty}\overline{A}_{o}\left(x,y\right)\exp\left[i2\pi\left(x\xi+y\eta\right)\Big/\lambda Z\right]dx\,dy\,,$$

$$(4.2.9)$$

which requires computation of the integral Fourier transform:

$$\overline{\Gamma}_{F}\left(\xi,\eta\right)=\int\limits_{-\infty}^{\infty}\int\limits_{-\infty}^{\infty}\overline{A}_{o}\left(x,y\right)\exp\left[i2\pi\left(x\xi+y\eta\right)\Big/\lambda Z\right]dx\,dy\;.\qquad(4.2.10)$$

Holograms synthesized according to Eq. 4.2.10 will be referred to as synthesized ***Fourier holograms***. Fourier and Fresnel holograms can be computed through Discrete Fourier and Fresnel transforms discussed in Ch. 3.

Schemes of using computer generated display Fresnel and Fourier holograms for visual observation of virtual objects are sketched in Fig. 4.7, a) and b), respectively.

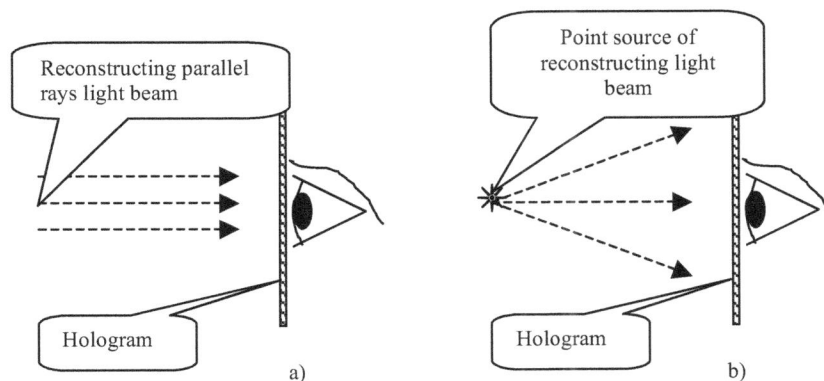

**Fig. 4.7.** Schematic diagram of visual observation of virtual objects using computer generated Fresnel (a) and Fourier (b) holograms

From the standpoint of object wave front reconstruction, Fresnel holograms differ from Fourier holograms in that they have focusing properties and are capable of reproducing the finite distance from the observation surface to the object. For reconstruction of Fourier holograms, spherical reconstruction light beam is needed and reconstructed image is observed in the plane of the reconstructing beam point source. In both cases optical transforms needed for hologram reconstruction are performed by the eye optics.

Having passed from the spatial 3D-problem to the problem of treating its 2D approximation, we, strictly speaking, have lost the possibility of taking into account the precise effect of object depth relief on the wave front. Even Fresnel holograms involve only a single distance value from the object to the observation plane rather than object relief depth. Nevertheless, it is still possible to synthesize holograms, which are capable of reconstructing 3-D images for visual observation and retain the most important property of holographic visualization - naturalness of object visual observation. At least two options exist for this: composite computer generated stereo- and macro-holograms and programmed diffuser holograms ([4]).

Stereo holograms are holograms synthesized for viewing by left and right eye. They reconstruct corresponding stereoscopic views of the object. Composite stereo holograms are composed of multiple holograms that reproduce different aspects of the object as observed from different horizontal positions. Macro-holograms are large composite stereo-holograms that cover wide observation angle in both horizontal and vertical directions and can be used as a wide observation window.

Programmed diffuser holograms are computer generated Fourier holograms that imitate properties of diffuse surfaces to scatter irradiation non-uniformly in the space and in this way provide visual clue about the object surface shape. Synthesis of stereo holograms and programmed diffuser hologram is detailed in Ch. 8

## 4.3    Digital-to-analog conversion problem of recording CGHs and spatial light modulators for recording CGHs

Optical media for recording computer-generated holograms may be classified into three categories: *amplitude-only media*, *phase-only media*, and *combined amplitude/phase media* (Fig. 4.8.).

In amplitude-only media, the controlled optical parameter is the light intensity transmission or reflection factor of the medium. Typical representatives of the amplitude media are the standard silver-halid photographic emulsions used in photography and optical holography (see for instance [8]). Photographic media have very good optical properties, such as resolution and dynamic range. However they have substantial drawbacks: they require wet chemical development, are not reversible and require electronic-to-optical conversion for hologram recording using electrical computer controlled signals. Recently, micro-lens array and micro-mirror array technology has emerged and reports were published on using them as amplitude media for recording computer generated holograms ([9,10]).

In phase-only media, optical thickness of the medium can be controllable, for example, by varying the medium refractive index, or its physical thickness, or both.

Phase-only media include thermoplastic materials, photo resists, bleached photographic materials, media based on photo polymers, liquid crystal media, etc.

Combined media allow independent control of both the light intensity transmission or reflection factor and of the optical thickness. Currently, these are photographic materials with two or more layers sensitive to radiation of different wavelengths. This permits the user to control of the transparency of certain layers and the optical thickness of others by exposing each layer to its wavelength independently.

Special digitally controlled hologram recording devices are required for modulating optical parameters of these media according to the mathematical hologram. No such special purpose devices are available as yet, and computer printer/plotters, display devices and other improvised means such as photo- and e-beam (electron) lithography are used instead.

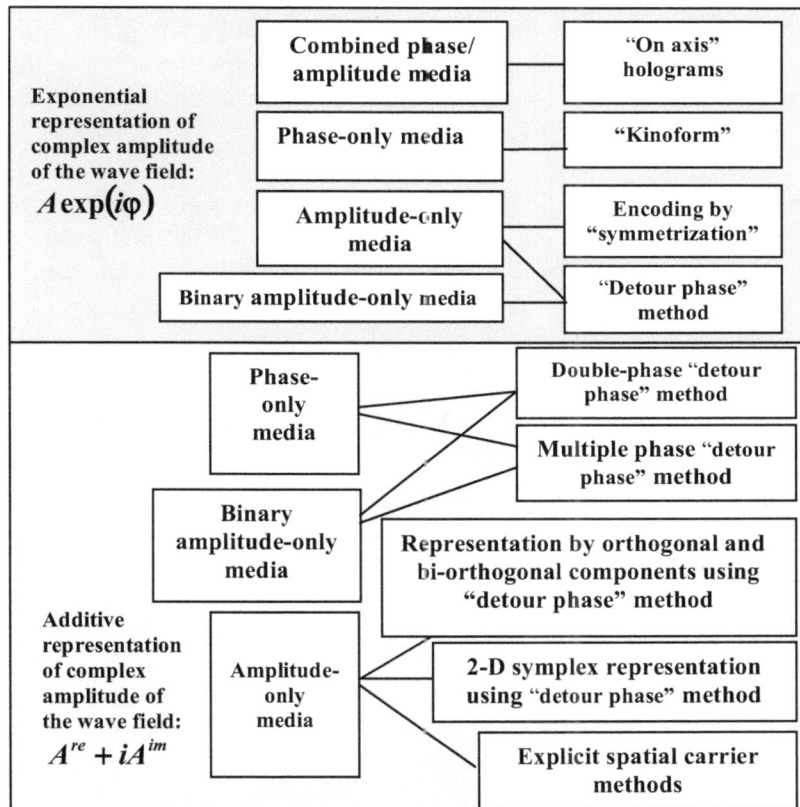

**Fig. 4.8.** Classification of methods for recording computer generated holograms

The distinctive feature of many of such devices is that they perform only binary, or two levels, modulation of the medium's optical parameters. Amplitude-only and phase-only media whose controlled optical parameters can assume only two values will be referred to as ***binary media***. The use of amplitude and phase media in the binary mode is quite inefficient in terms of the medium information capacity because the possibility of writing information on them is defined only by their spatial degrees of freedom (their resolution power), whereas in amplitude and phase media, in principle, the degrees of freedom related to a transmission (reflection, refraction) factor dynamic range can be used as well. The major advantage of the binary mode is the simplicity of media exposure and copying recorded holograms.

The most important numerical characteristics of hologram recorders are the total number of cells, which may be exposed, and their sampling interval, that is, the distances $\Delta\xi$ and $\Delta\eta$ between neighboring, separately and independently exposed resolution cells. The sampling interval defines the angular dimensions $\left(\theta_x, \theta_y\right)$ of the reconstructed image according to the known diffraction relationships:

$$\theta_x = 2\pi\lambda/\Delta\xi; \ \theta_y = 2\pi\lambda/\Delta\eta, \qquad (4.2.11)$$

For instance, in order to make the reconstructed image's angular dimensions about $20^o$ or more with a reconstruction light wavelength of about 0.5 μm, $\Delta\xi$ and $\Delta\eta$ should be less then 10 μm. Existing spatial light modulators and printing devices that can be used for recording computer generated holograms have a sampling step of about 1 through 10 μm, and the total number of resolution cells, by the order of magnitude, from $10^3 \times 10^3$ to $10^4 \times 10^4$.

Table 1 presents, as an example, a short specification of one of modern spatial light modulators "Reflective Phase only display panel (LCoS - Liquid Crystal on Silicon)" (www.holoeye.com) shown in Fig. 4.9. Fig. 4.10 shows amplitude and phase modulation transfer functions of this SLM.

**Tab. 1:** Reference data for HDTV phase-only LCOS

| Active area dimensions | 15.36 x 8.64 (0.7" diagonal) mm |
|---|---|
| Screen aspect ratio | 16 (H) : 9 (V) |
| Display resolution | 1920 (H) x 1080 (V) pixels |
| Pixel pitch | 8.0 μm |
| Pixel configuration | Orthogonal |
| Phase Levels | 256 (8 bit) |
| Optical Mode | Reflective |
| Liquid crystal type | Nematic (ECB-mode) |

**Fig. 4.9.** Reflective Phase only LCoS display panel (adopted from http://www.holoeye.com )

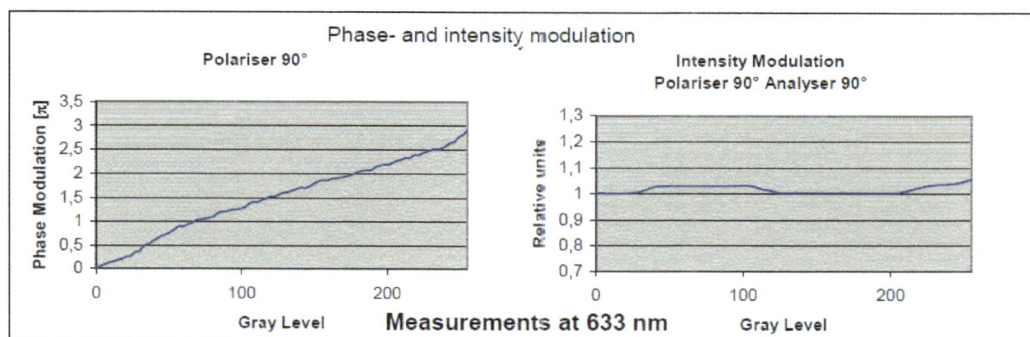

**Fig. 4.10.** Phase and intensity modulation transfer functions of the LCoS display panel (adopted from http://www.holoeye.com )

## References

1.    Brown B. R., Lohmann A., Complex spatial filtering with binary masks, *Appl. Optics*, **5**, No. 6, 967-969 (1966)

2.    Brown B. R., Lohmann A., Computer generated binary holograms, IMB J. Res. Dev., v. 13, No. 2, 160-168 (1969

3.    Kronrod M.A., Merzlyakov N.S., Yaroslavsky L.P., Computer Synthesis of Transparency Holograms, Soviet Physics-Technical Physics, v. 13, 1972, p. 414 - 418.

4.    Yaroslavskii L., Merzlyakov N., Methods of Digital Holography, Consultance Bureau, N.Y., 1980

5.    Lohmann A., A Prehistory of Computer Generated Holography, OPN, February, 2008

6.    Yaroslavsky L., Image input-output devices for digital computers, Gosenergonizdat, Moscow, 1968 (In Russian)

7.    Soifer V., Kotlyar V., Doskolovich L., Iterative methods for diffractive optical elements computation, Taylor & Francis Ltd, 1997

8.    Bjelkhagen H. I., Silver Halid Recording Materials, Springer Verlag, Heidelberg, 1993

9.    Nesbitt R.S., Smith S. L., Molnar R.A., Benton S. A., Holographic recording using a digital micromirror device, Conf. on Practical Holography, Proc. SPIE, 3637, 12-20, 1999

10.    Kreis T., Aswendt P., Hoeffling R., Holographic reconstruction using a digital micromirror device, Optical Engineering, v. 40 (6), pp. 926-933, June 2001

# 5    Hologram Encoding for Recording on Physical Media

**Abstract:** The purpose of hologram encoding is converting complex valued samples of mathematical holograms obtained in the result of synthesis of computer-generated holograms into data that can be used to directly control physical parameters of recording media. This chapter overviews different hologram encoding methods oriented on different recording media that can be used for this purpose.

## 5.1    Methods for hologram encoding for amplitude media.

### 5.1.1    Detour phase method

Historically, the first method for recording computer-generated holograms was proposed by A. Lohmann and his collaborators for amplitude-only binary media ([1,2]). In this method, illustrated in Fig. 5.1, individually controlled elementary cells of the medium are allocated, one cell per sample, for recording amplitudes and phases of samples of mathematical holograms. The modulus of the complex number representing the sample defines the size of the opening (aperture) in the cell and the phase - the position of the opening within the cell.

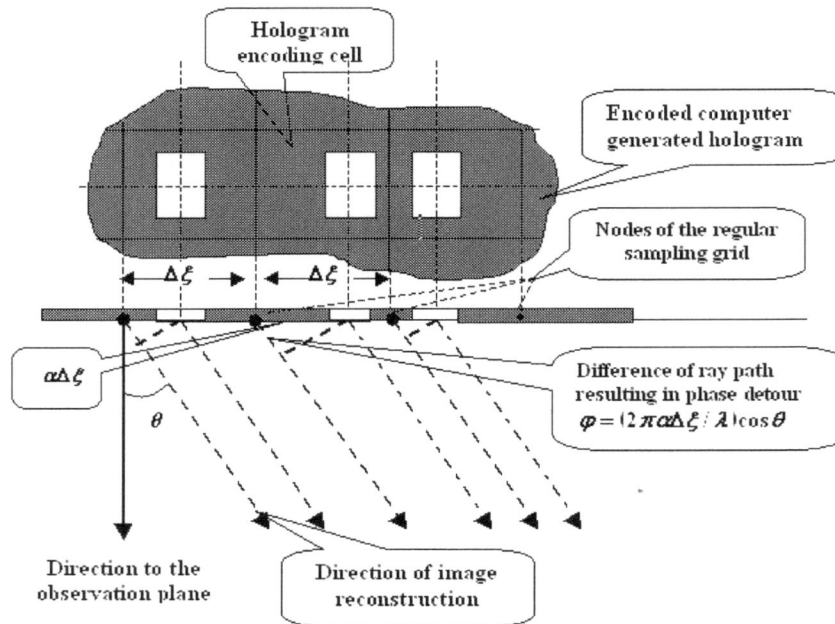

**Fig. 5.1.** Detour phase method for encoding phase shift by a spatial shift of the transparent aperture

All cells are arranged in a rectangular lattice, or *sampling raster*. A shift of the aperture by $\Delta\xi$ in a given cell with respect to its raster node corresponds to a phase detour for this cell equal to $(2\pi\Delta\xi\cos\theta)/\lambda$ for hologram reconstruction at an angle $\theta$ to the system's optical axis perpendicular to the hologram plane, where $\lambda$ is wavelength of light used for hologram reconstruction. The use of the aperture spatial shifts for representing phases of complex numbers is known as the *detour phase method*.

By default, samples of the mathematical hologram are defined for nodes of the hologram sampling raster. In order to determine, for each sample of the mathematical hologram, exact position and size of the opening for encoding the phase and amplitude of this sample, one should, in principle, have values of phase and amplitude of the mathematical hologram in all positions between given samples. This requires generating a continuous model of the mathematical hologram by means of interpolation of its available samples. Interpolated amplitude and phase profiles can be re-sampled in positions, in which phase detour of the opening for each hologram recording cell coincides with the phase profile of the mathematical hologram as it is illustrated in Fig. 5. 2.

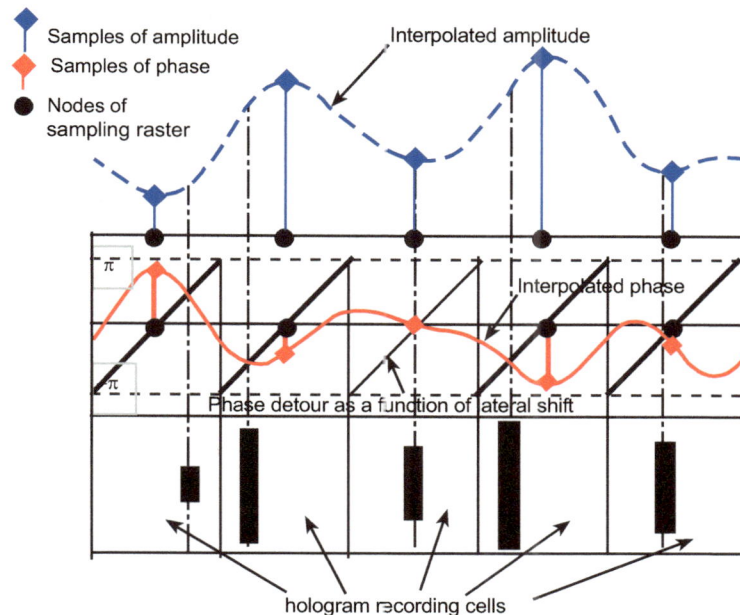

**Fig. 5.2.** Illustration of hologram re-sampling in phase-detour method of hologram encoding. Black rectangles are openings placed in positions, where their phase detour coincides with the interpolated phase profile of the mathematical hologram

For interpolation, different methods can be used ([3]). The simplest one, the nearest neighbor interpolation, is the fastest and the least accurate. It results in considerable distortions in reconstructed images. Some of them will be illustrated in Chapt. 7, Sect 7.4. Brown and Lohmann ([1]) tried in their experiments the Newton's interpolation method using four neighboring sampling points, which did show certain improvement of reconstructed images, though the need for further improvement remained. The most accurate interpolation method for sampled data is the discrete sinc-interpolation ([3]). More accurate resampling requires more computations. Specifically, with perfect discrete sinc-interpolation, the number of operations for computing interpolated mathematical hologram is equal to the number of operations for computing non-interpolated sampled hologram times the number of intermediate interpolated samples per each initial one. This is a computational drawback of the method.

The method also quite inefficiently utilizes spatial resolution of the recording media. As was already noted, only spatial degrees of freedom are used for recording on binary media; therefore, the number of binary medium cells should exceed the number of hologram samples by a factor equal to the product of the number the amplitude and phase quantization levels. This product may run into several tens or even hundreds. Such low efficiency in using degrees of freedom of the recording media is probably, a major drawback of binary holograms. However many modern technologies such as lithography and laser and electron beam pattern generators provide much higher number of spatial degrees of freedom that the required number of hologram samples. Therefore this drawback is frequently disregarded and such merits of binary recording as simpler recording and copying technology of synthesized holograms and the possibility of using commercially available devices made it the most widespread. A number of modifications of the method are known, oriented to different types of recording devices ([4]).

### 5.1.2   *Orthogonal, bi-orthogonal and 2-D simplex encoding methods*

In an additive representation, complex numbers are regarded as vectors on the complex plane in the orthogonal coordinate system representing real and imaginary parts of the numbers as it is shown in Fig. 5.3, a).

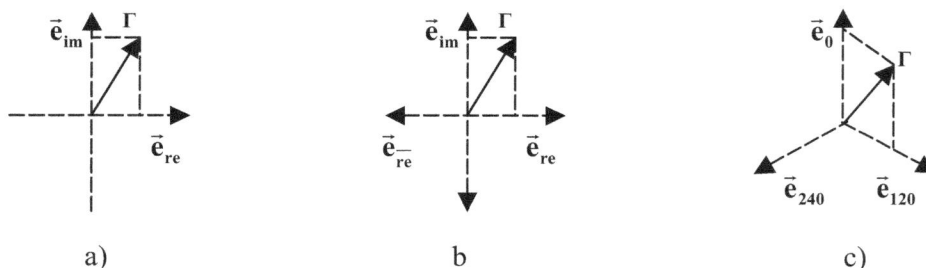

**Fig. 5.3.** Additive representation of complex numbers in orthogonal basis (a), in biorthogonal basis (b) and in 2-D simplex (c).

The representation of a vector $\tilde{\mathbf{A}}$ by its real $\tilde{A}_{re}$ and imaginary $\tilde{A}_{im}$ parts:

$$\tilde{A} = \tilde{A}_{re}\mathbf{e}_{re} + \tilde{A}_{im}\mathbf{e}_{im}, \tag{5.1.1}$$

where $\mathbf{e}_{re}$ and $\mathbf{e}_{im}$ are orthogonal unit vectors, is the base of the hologram encoding method, which we will refer to as the ***orthogonal encoding method***. When recording a hologram, the phase angle between the orthogonal components, which is equal to $\pi/2$, can be encoded by the detour phase method by means of recording $\tilde{A}_{re}$ and $\tilde{A}_{im}$ into neighboring hologram resolution cells in sampling raster rows as it is shown in Fig.5.4, a). The reconstructed image will be in this case observed at an angle $\theta_\xi$ to this axis defined by the equation

$$\Delta\xi\cos\theta_\xi = \lambda/4. \tag{5.1.2}$$

For recording negative values of $\tilde{A}_{re}$ and $\tilde{A}_{im}$, a constant positive bias can be added to recorded values.

With such encoding and recording of holograms, one must take into account that the optical path differences $\lambda/2$ and $3\lambda/4$ in the same direction under angle $\boldsymbol{\theta_\xi}$ will correspond to the next pair of hologram resolution cells (see Fig.5.3, a). This implies that values of $\tilde{A}_{re}$ and $\tilde{A}_{im}$ for each second sample of the mathematical hologram should be recorded with opposite signs.

Since two resolution elements are used here for recording one hologram sample, recorded holograms have a double redundancy with respect to the hologram recording cells.

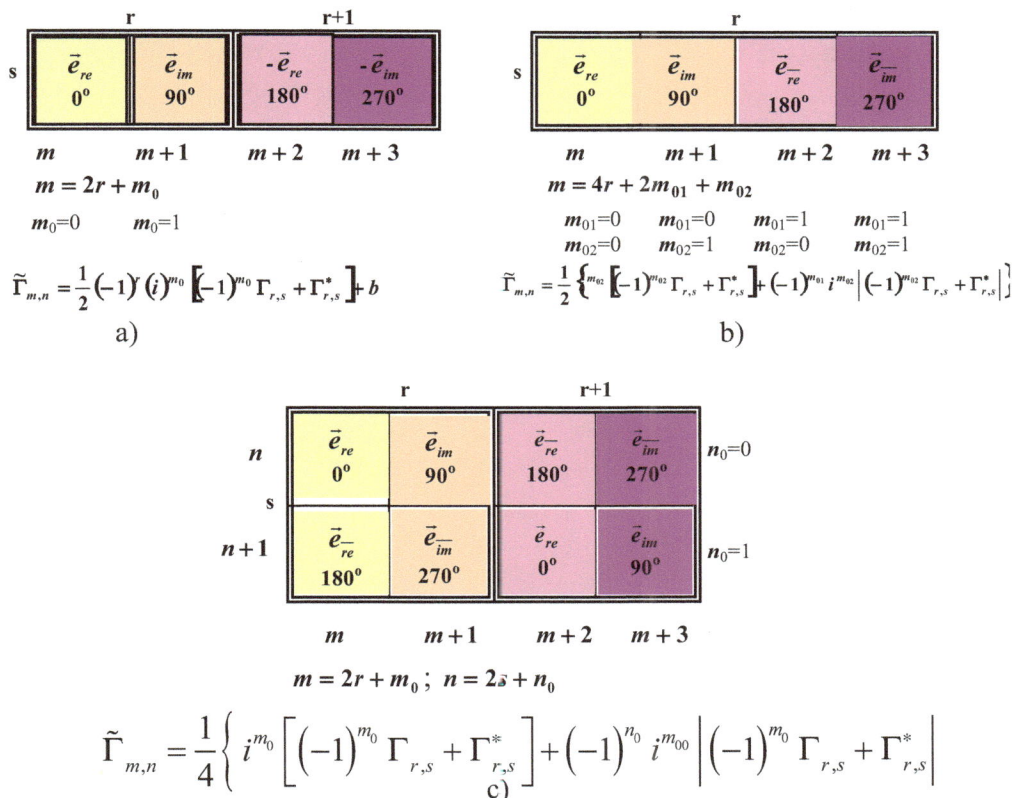

**Fig. 5.4.** Hologram encoding by decomposition of complex numbers in orthogonal (a) and biorthogonal (b, c) bases. Groups of hologram encoding cells corresponding to one hologram sample are outlined with a double line. Numbers in the cells indicate phase detour angle that corresponds to the cell.

The orthogonal encoding technique can be described formally as follows. Let $\{r,s\}$ be indices of samples $\{\Gamma_{r,s}\}$ of the mathematical hologram, $r = 0,1,...,M-1$, $\{m,n\}$ be indices of the recording medium resolution cells, $\{\tilde{\Gamma}_{r,s}\}$ be samples of the encoded hologram ready for recording. Let also index $m$ be counted as a two digits number:

$$m = 2r + m_0, \quad m_0 = 0,1, \tag{5.1.3}$$

and index $n$ be counted as $n = s$. The encoded hologram can then be written as

$$\tilde{\Gamma}_{m,n} = \tilde{\Gamma}_{2r+m_0,s} = \frac{1}{2}(-1)^r (i)^{m_0} \left[ (-1)^{m_0} \Gamma_{r,s} + \Gamma_{r,s}^* \right] + b, \qquad (5.1.4)$$

where $b$ is a positive bias constant needed to secure that recording values are non-negative.

In reconstruction of computer generated holograms recorded with a constant bias, substantial part of energy of the reconstructing light beam is not used for image reconstruction and goes to the zero-order diffraction spot. One can avoid the constant biasing for recording $\Gamma_{re}$ and $\Gamma_{im}$ by allocating, for recording one hologram sample, four neighboring in sampling raste medium resolution cells rather then two. This arrangement is shown in Fig.5.4, b). With a reconstruction angle defined by Eq. (5.1.2), phase detours $0$, $\pi/2$, $\pi$ and $3\pi/2$ will correspond to them. Therefore, cells in each group of four cells allocated for recording of one sample of the mathematical hologram should be written in the following order: $\left(\Gamma_{re} + |\Gamma_{re}|\right)/2$; $\left(\Gamma_{im} + |\Gamma_{im}|\right)/2$; $\left(|\Gamma_{re}| - \Gamma_{re}\right)/2$; $\left(|\Gamma_{im}| - \Gamma_{im}\right)/2$. One can see that in this case all recorded values are nonnegative. This method represents a vector in the complex plane in a **biorthogonal basis** ($\vec{e}_{re}, \vec{e}_{\overline{re}}, \vec{e}_{im}, \vec{e}_{\overline{im}}$) shown in Fig. 5.3, b):

$$\tilde{A} = \frac{1}{2}\left\{ \left(\Gamma_{re} + |\Gamma_{re}|\right)\vec{e}_{re} + \left(\Gamma_{re} - |\Gamma_{re}|\right)\vec{e}_{\overline{re}} + \left(\Gamma_{im} + |\Gamma_{im}|\right)\vec{e}_{im} + \left(\Gamma_{im} - |\Gamma_{im}|\right)\vec{e}_{\overline{im}} \right\}. \quad (5.1.5)$$

In terms of indices $(m,n)$ of recording medium resolution cells counted as

$$m = 4r + 2m_{01} + m_{02} ; r = 0,1,..., M-1; \ m_{01} = 0,1; \ m_{02} = 0,1; n = s, \quad (5.1.6)$$

this encoding method can be described as follows:

$$\tilde{\Gamma}_{m,n} = \tilde{\Gamma}_{4r+2m_{01}+m_{02},s} =$$
$$\frac{1}{2}\left\{ i^{m_{02}} \left[ (-1)^{m_{02}} \Gamma_{r,s} + \Gamma_{r,s}^* \right] + (-1)^{m_{01}} i^{m_{02}} \left| (-1)^{m_{02}} \Gamma_{r,s} + \Gamma_{r,s}^* \right| \right\} \qquad (5.1.7)$$

In hologram encoding by this method, it is required, that the size of the medium resolution cell in one direction be four times smaller than that in the perpendicular one in order to preserve proportions of the reconstructed picture. A natural way to avoid this anisotropy is to allocate pairs of resolution cells in two neighboring raster rows for each sample of the mathematical hologram as it is shown in Fig.5.4, c). In this case encoded hologram samples $\{\tilde{\Gamma}_{m,n}\}$ are recorded according to the following relationship:

$$\tilde{\Gamma}_{m,n} = \tilde{\Gamma}_{2r+m_0,2s+n_0} = \frac{1}{4}\left\{ i^{m_0} \left[ (-1)^{m_0} \Gamma_{r,s} + \Gamma_{r,s}^* \right] + (-1)^{n_0} i^{m_{00}} \left| (-1)^{m_0} \Gamma_{r,s} + \Gamma_{r,s}^* \right| \right\}, \quad (5.1.8)$$

where indices $(m,n)$ are counted as two digits numbers:

$$m = 2r + m_0; \; n = 2s + n_0; \; m_0 = 0,1; \; n_0 = 0,1. \tag{5.1.9}$$

With this encoding method, images are reconstructed along a direction making angles $\theta_\xi$ and $\theta_\eta$ with axes $(\xi, \eta)$ in the hologram plane defined by equations:

$$\Delta\xi \cos\theta_\xi = \lambda / 2; \; \Delta\eta \cos\theta_\eta = \lambda / 2. \tag{5.1.10}$$

Representation of complex numbers in the bi-orthogonal basis is redundant because two of four components are always zero. This redundancy is reduced in the representation of complex numbers with respect to a **two-dimensional simplex** ($\vec{e}_0, \vec{e}_{120}, \vec{e}_{240}$):

$$\tilde{A} = \Gamma_0 \vec{e}_0 + \Gamma_1 \vec{e}_{120} + \Gamma_2 \vec{e}_{240}. \tag{5.1.11}$$

as it is illustrated in Fig.5.3, c). Similarly to the bi-orthogonal basis, this basis is also redundant because

$$\vec{e}_0 + \vec{e}_{120} + \vec{e}_{240} = 0. \tag{5.1.12}$$

This redundancy is exploited to insure that components $\{\Gamma_0, \Gamma_1, \Gamma_2\}$ of complex vectors be non-negative. Formulas for computing these components are presented, along with diagrams illustrating their derivation, in Table 5.1.

**Table 5.1.**

| | | |
|---|---|---|
| $\Gamma_0 = \left\|\Gamma^{im}\right\| + \dfrac{\sqrt{3}}{4}\left\|\Gamma^{re}\right\|$ | $\Gamma_0 = \left\|\Gamma^{im}\right\| + \dfrac{\sqrt{3}}{4}\left\|\Gamma^{re}\right\|$ | $\Gamma_0 = 0$ |
| $\Gamma_1 = \dfrac{\sqrt{3}}{2}\left\|\Gamma^{re}\right\|$ | $\Gamma_1 = 0$ | $\Gamma_1 = \left\|\Gamma^{im}\right\| - \dfrac{\sqrt{3}}{4}\left\|\Gamma^{re}\right\|$ |
| $\Gamma_2 = 0$ | $\Gamma_0 = \dfrac{\sqrt{3}}{2}\left\|\Gamma^{re}\right\|$ | $\Gamma_2 = \Gamma_1 + \dfrac{\sqrt{3}}{2}\left\|\Gamma^{re}\right\| = \left\|\Gamma^{im}\right\| + \dfrac{\sqrt{3}}{4}\left\|\Gamma^{re}\right\|$ |

When recording holograms with this method, one can encode the phase angle corresponding to unit vectors ($\vec{e}_0, \vec{e}_{120}, \vec{e}_{240}$) by means of the above-mentioned detour phase method by writing the components $\{\Gamma_0, \Gamma_1, \Gamma_2\}$ of the simplex-decomposed vector into groups of three neighboring hologram resolution cells in the raster as it is shown in Fig.5.5 , a), b) and c)). The latter two arrangements are more isotropic than the first one. Coordinate scale ratio for arrangement b) is 2:1.5 and for arrangement c) is $3\sqrt{3}:4$ whereas for arrangement a) it is 3:1.

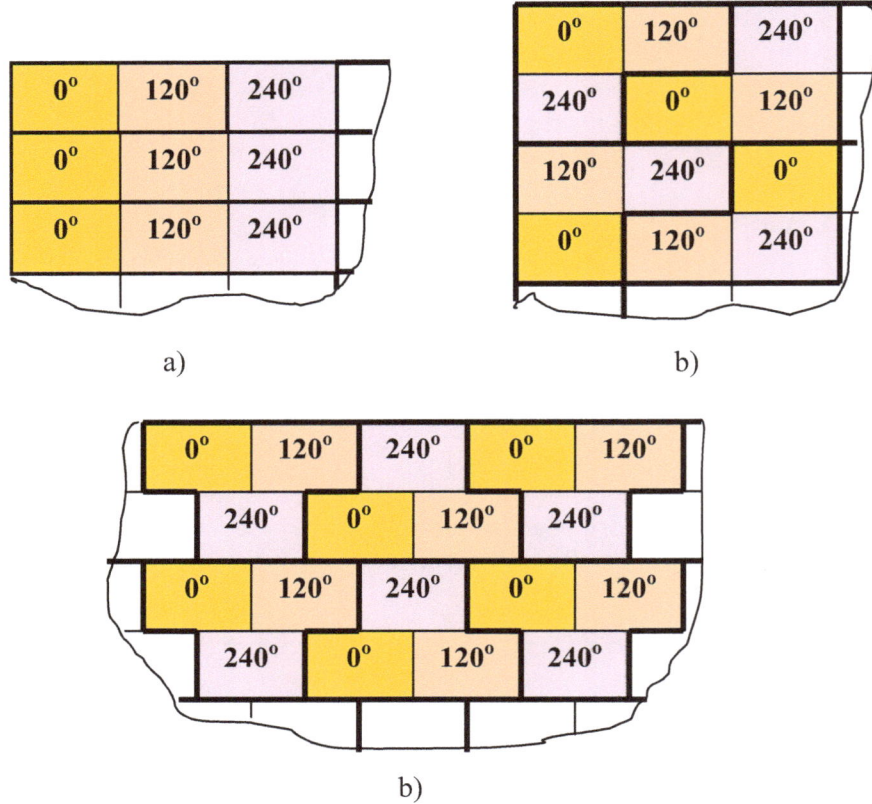

a)                                                                         b)

b)

**Fig. 5.5.** Three methods of arrangement of recording medium resolution cells for hologram encoding by decomposition of complex numbers with respect to a 2-D simplex. Triples of medium resolution cells used for recording one hologram sample are outlined by bold lines.

For holograms recorded by this method according to the arrangement of Fig. 5.5, a), images are reconstructed at angles $\theta_\xi = \arccos(\lambda / 3\Delta\xi)$ to the axis $\xi$ coinciding with the direction of the hologram raster rows and $\theta_\eta = 0$ to the perpendicular axis. For the arrangements of Fig. 5.5, b) and c) the reconstruction angles are, respectively, $\theta_\xi = \arccos(\lambda / 3\Delta\xi)$, $\theta_\eta = \arccos(2\lambda / 3)$ and $\theta_\xi = \arccos(4\lambda / 3\Delta\xi)$, $\theta_\eta = \arccos(2\lambda / 3)$.

Above-described encoding methods based on additive representation of complex vectors are applicable for recording on amplitude-only media, both continuous tone and binary. In the latter case, coded hologram components, which are projections on basic vectors of complex numbers representing hologram samples, are recorded by varying the size of the transparent aperture in each of the appropriate resolution cells.

### 5.1.3  *"Symmetrization" method*

In hologram recording on amplitude media, the main problem is recording the phase component of hologram samples. The **symmetrization method** ([5]) offers a straightforward solution of this problem for recording Fourier holograms. The method assumes that, prior to hologram synthesis, the object is symmetrized, so that, according to properties of the Discrete Fourier transform, its Fourier hologram contains only real valued samples and, thus, can be recorded on amplitude-only media. As real numbers may be both positive and negative, holograms should be recorded on amplitude-only media with a constant positive bias, making all the recorded values positive.

For Shifted DFT as a discrete representation of the integral Fourier transform, symmetry properties require symmetrization through a rule that depends on the shift parameters $(u,v)$. For example, for integer $2u$ and $2v$, the symmetrization rule for the object wave front specified by the array of its samples $\left\{\overline{A}_{k,l}\right\}$, $k = 0,1,...,N_1 - 1$, $l = 0,1,...,N_2 - 1$ is:

$$\overset{\sim}{\overline{A}}_{k,l} = \begin{cases} \overline{A}_{k,l}, 0 \le k \le N_1 - 1;\ 0 \le l \le N_2 - 1 \\ \overline{A}_{2N-k,l},\ N_1 \le k \le 2N_1 - 1;\ 0 \le l \le N_2 - 1 \end{cases} . \qquad (5.1.13)$$

In doing so, the number of samples of the object and, correspondingly, of its Fourier hologram, is twice that of the original object. It is this two-fold redundancy that enables one to avoid recording the phase component. We refer to this symmetrization method as **symmetrization by duplication**. It is illustrated in Fig. 5.6, a).

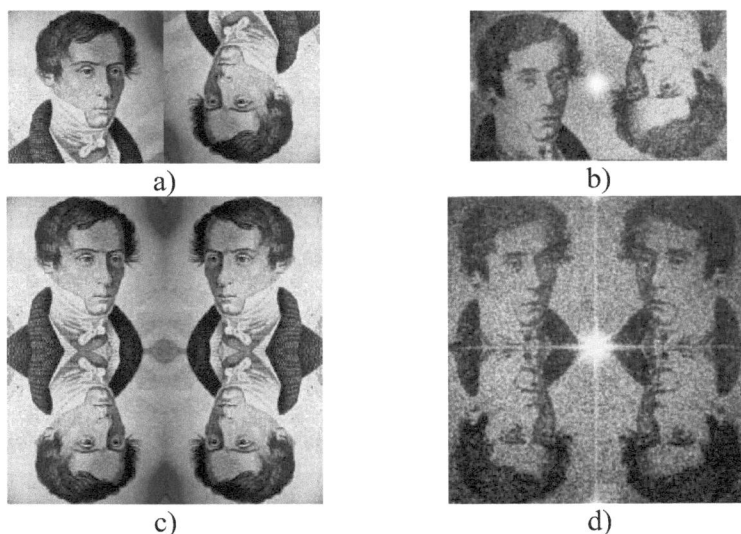

a)     b)

c)     d)

**Fig. 5.6.** Symmetrization of images for recording computer-generated holograms on amplitude media: a), b) – symmetrization by duplication: symmetrized images and an example of an image optically reconstructed from computer-generated hologram; c), d) the same for symmetrization by quadruplicating.

***Symmetrization by quadruplicating*** is possible as well. It consists in symmetrizing the object according to the rule of Eq. 5.1.13 with respect to both indices ***k*** and ***l*** as it is illustrated in Fig. 5.6, b). Hologram redundancy becomes in this case four-fold.

Holograms of symmetrized objects are also symmetrical and reconstruct duplicated or quadruplicated objects depending on the particular symmetrization method (Figs. 5.6, c) and d)). Note that duplicating and quadruplicating do not necessarily imply a corresponding increase in computation time for execution of SDFT at the hologram calculation step because, for computing SDFT, one can use combined algorithms making use of signal redundancy for accelerating computations ([3, 6]).

## 5.2   Methods for recording computer-generated holograms on phase media.

### 5.2.1   Double-phase and multiple-phase methods

For recording on phase media, additive representation of complex vectors assumes complex vector representation as a sum of complex vectors of a standard length. Two versions of such encoding are known: double phase method and multiple phase method.

In the ***double-phase method*** ([7]), hologram samples are encoded as

$$\tilde{A} = |\Gamma|\exp(i\phi) = A_0\exp(i\phi_1) + A_0\exp(i\phi_2),\tag{5.2.1}$$

where $(\varphi_1, \varphi_2)$ are component phase angles defined by the following equations:

$$\varphi_1 = \varphi + \arccos(|\Gamma|/2A_0)\ ;\ \varphi_2 = \varphi - \arccos(|\Gamma|/2A_0),\tag{5.2.2}$$

that can be easily derived from the geometry of the method illustrated in Fig. 5. 7.

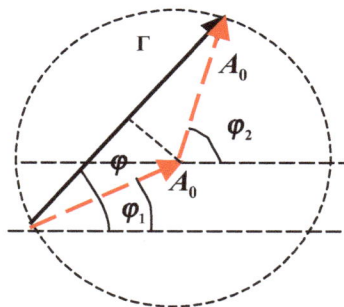

**Fig. 5.7.** Representation of a complex vector as a sum of two vectors of equal length

Two neighboring medium resolution cells should be allocated for representing two vector components. Formally, the encoded in this way hologram can be written as

$$\tilde{\Gamma}_{m,n} = \tilde{\Gamma}_{2r+m_0,s} = A_0\exp\left\{i\left[\varphi_{r,s} + (-1)^{m_0}\arccos(|\Gamma_{r,s}|/2A_0)\right]\right\},\tag{5.2.3}$$

where $\left\{ \tilde{\mathbf{A}}_{r,s} = \left| \tilde{\mathbf{A}}_{r,s} \right| \exp\left( i\phi_{r,s} \right) \right\}$ are samples of the mathematical hologram, indices of the recording medium resolution cells $(m,n)$ are counted as $m = 2r + m_0$, $m_0 = 0,1$, $n = s$ and $A_0$ is found as half of maximal value of $\left\{ \left| \tilde{\mathbf{A}}_{r,s} \right| \right\}$.

For hologram encoding with the double-phase method and recorded on phase media, images are reconstructed in the direction normal to the hologram plane, because the optical path difference of rays passing along this direction through the neighboring resolution cells is zero. This holds, however, only for the central area of the image through which the optical axis of the reconstruction system passes. In the peripheral areas of the image, some nonzero phase shift between the rays appears, thus leading to distortions in the peripheral image. This phenomenon will be discussed in more details in Chapt. 7.

The double phase method can also be used for recording on amplitude binary media with phase-detour encoding of phases. Two versions of such an implementation of the double-phase method were suggested ([7]). In the first version, two elementary cells of the binary medium are allocated to each of the two component vectors, their phases being coded by a shift of the transparent (or completely reflecting) aperture along a direction perpendicular to the line connecting cell centers as it is illustrated in Fig.5.8, a). This technique exhibits above-mentioned distortions due to mutual spatial shift of elementary cells. The second version reduces these distortions by means of decomposing each elementary cell into sub-cells alternating as it is shown in Fig.5.8, b).

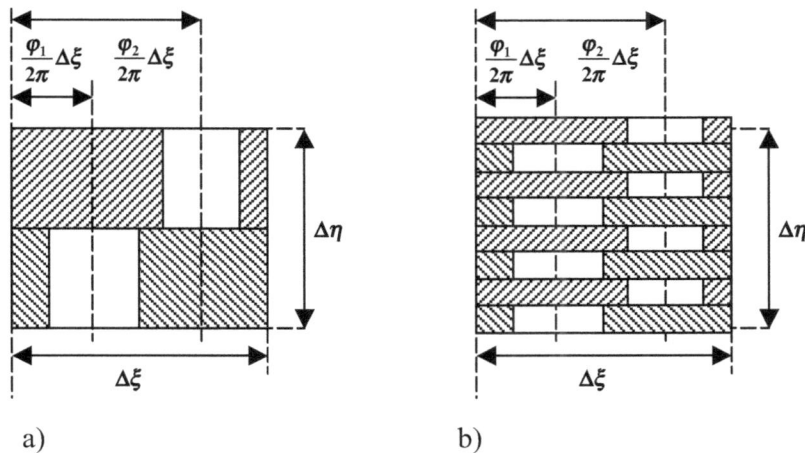

a)                                           b)

**Fig. 5.8.** Double-phase encoding method for hologram recording on a binary phase medium: two separate cells (a); decomposition of cells into alternating sub-cells

The double-phase method is readily generalized to ***multi-phase encoding*** by vector decomposition into an arbitrary number $Q$ of equal-length vector components:

$$\tilde{\mathbf{A}} = \sum_{q=1}^{Q} A_0 \exp\left(i\phi_q\right).$$  (5.2.4)

As $\Gamma$ is a complex number, Eq.(5.2.4) represents two equations for $Q$ unknown values of $\{\varphi_q\}$. They have a unique solution defined by Eq. 5.2.2 only for $Q = 2$. When $Q > 2$, $\{\varphi_q\}$ can be chosen in a rather arbitrary manner. For example, for odd $Q$ one can choose phases $\{\varphi_q\}$ making an arithmetic progression:

$$\varphi_{q+1} - \varphi_q = \varphi_q - \varphi_{q-1} = \Delta\varphi .$$  (5.2.5)

In this case, the following equations can be derived for the increment $\Delta\varphi$:

$$\frac{\sin\left(Q\Delta\varphi / 2\right)}{\sin\left(\Delta\varphi / 2\right)} = \frac{|\Gamma|}{A_0} .$$  (5.2.6)

and for phase angles $\{\varphi_q\}$:

$$\varphi_q = \varphi + \left[q - \left(Q+1\right)/2\right]\Delta\varphi , \quad q = 1, 2, ..., Q .$$  (5.2.7)

Eq. 5.2.6 reduces to an algebraic equation of power $\left(Q-1\right)/2$ with respect to $\sin^2\left(\Delta\varphi / 2\right)$. Thus, for $Q = 3$,

$$\Delta\varphi = 2\arcsin\left(\frac{\sqrt{3}}{2}\sqrt{1 - |\Gamma|/3}\right).$$  (5.2.8)

For even $Q$, it is more expedient to separate all the component vectors into two groups having the same phase angles $\varphi_1$ and $\varphi_2$ that are defined by analogy with Eq.(5.2.2) as

$$\varphi_+ = \varphi + \arccos\left(|\Gamma|/QA_0\right); \quad \varphi_- = \varphi - \arccos\left(|\Gamma|/QA_0\right).$$  (5.2.9)

With multi-phase encoding, the dynamic range of possible hologram values may be extended because the maximal reproducible amplitude in this case is $QA_0$. Most interesting of the $Q > 2$ cases are those of $Q = 3$ and $Q = 4$ because the two-dimensional spatial degrees of freedom of the medium and hologram recorder can be used more efficiently through allocation of the component vectors according to Figs.5.4, c) and 5.5, b) and c).

## 5.2.2   Kinoform

One of the most popular encoding methods oriented on using phase-only media is that of *kinoform* ([8]). Kinoform is a computer-generated hologram in which variations of amplitude data of the mathematical hologram are disregarded, amplitudes of all hologram samples are set to a constant and only sample phases are recorded on a phase-only medium. It is used, when approximative reconstruction of only the amplitude component of the reconstructed object wave front is required in hologram reconstruction.

In synthesis of kinoform, a specially prepared array of pseudo-random numbers in the range $\left(-\pi,\pi\right)$ is assigned for the phase component of the object wave front. The array should be generated in such a way as to reconstruct, from the kinoform, in the Fourier plane of an optical reconstruction setup a certain given spatial distribution of light intensity that reproduces the amplitude of the object wave front.

Of course, disregarding variations of hologram sample amplitudes results in distortions of the hologram which manifest themselves in appearance of speckle noise in the reconstructed wave front. However, kinoforms are advantageous in terms of the use of energy of the reconstruction light because the total energy of the reconstruction light is transformed into the energy of the reconstructed wave field without being absorbed by the hologram medium. Moreover, the distortions can to some degree be reduced by an appropriate choice of artificially introduced phase component of the object wave field using the following iterative optimization procedure ([10]).

Let $\left\{\vartheta_{k,l}\right\}$ be an array of numbers in the range $\left(-\pi,\pi\right)$ representing the required phase component of the object wave front and $\left\{\left|A\left(k,l\right)\right|\right\}$ be samples of the object wave front amplitude, where $\left\{k,l\right\}$ are samples' indices. This phase component must satisfy the equation

$$\left|A\left(k,l\right)\right| = C\cdot abs\left\{\mathrm{IDFT}\left(\exp\left(\mathrm{angle}\left\{\mathrm{DFT}\left\{\left|A\left(k,l\right)\right|\exp\left(i\vartheta_{k,l}\right)\right\}\right\}\right)\right)\right\} \qquad (5.2.10)$$

where $C$ is an appropriate normalizing constant, $abs\left(.\right)$ is an operator of taking absolute value of the variable and $\mathrm{angle}\left(\cdot\right)$ is an operator for computing the phase component of the complex variable. In general, the exact solution of the equation may not exist. However, one can always replace it by a solution $\left\{\hat{\vartheta}_{k,l}\right\}$, taken from a certain class of phase distributions, that minimizes an appropriate measure $D\left(\cdot,\cdot\right)$ of deviation      of      the      result      of      numerical      reconstruction $C\cdot abs\left\{\mathrm{IDFT}\left(\exp\left(\mathrm{angle}\left\{\mathrm{DFT}\left\{\left|A\left(k,l\right)\right|\exp\left(i\vartheta_{k,l}\right)\right\}\right\}\right)\right)\right\}$   of   the   kinoform   from $\left\{\left|A\left(k,l\right)\right|\right\}$:

$$\left\{\hat{\vartheta}_{k,l}\right\} = \underset{\left\{\vartheta_{k,l}\right\}}{\arg\min}\left[D\left(\left\{\left|A(r,s)\right|\right\}, C \cdot abs\left\{\mathrm{IDFT}\left(\exp\left(\mathrm{angle}\left\{\mathrm{DFT}\left\{\left|A(k,l)\right|\exp\left(i\vartheta_{k,l}\right)\right\}\right\}\right)\right)\right\}\right)\right]$$

,                                                                                  (5.2.11)

where Pseudo-random number generator can serve as a source for realizations of the phase distributions and a solution of Eq. 5.2.11 can be searched through an iterative procedure according to a flow diagram shown in Fig. 5.9. One can also include in this process, when necessary, quantization of the phase distribution $\theta_{r,s}$ of the kinoform to a certain specified number of quantization levels (operator $\mathrm{quant}(\cdot)$ in the flow diagram) in order to take into account properties of the phase spatial light modulator intended for recording of the kinoform

Flow diagram of Fig. 5.9 assumes that iterations are carried out over one realization of the array of primary pseudo-random numbers. In principle, one can repeat the iterations for different realizations and then select of all obtained phase masks the one that provides the least deviation from the required distribution $\left\{\left|A(k,l)\right|\right\}$.

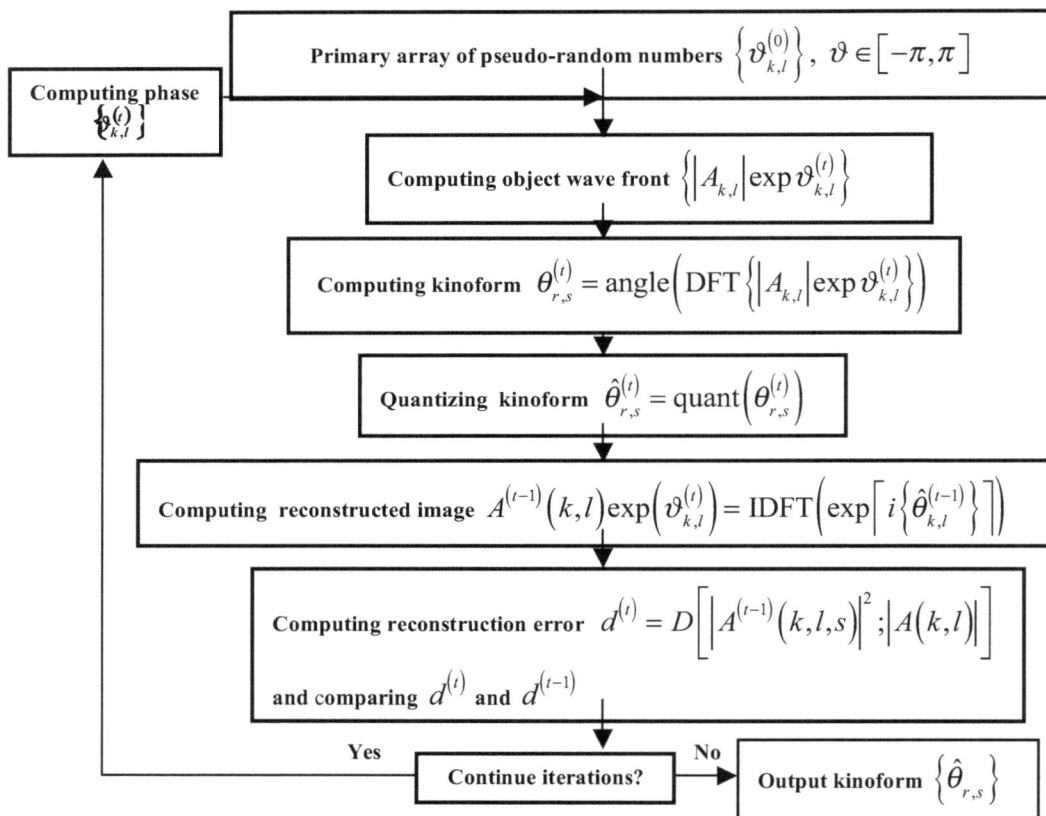

**Fig. 5.9.** Flow diagram of an iterative algorithm for generating pseudo-random phase mask with a specified power spectrum.

Illustrative examples of images reconstructed, in computer simulation, from kinoforms synthesized by the described method with 255 and 5 phase quantization levels are shown in Fig. 5.10, b) and c) along with original image (Fig. 5.10, a) used as the amplitude mask $\left[ A(k,l) \right]$. In this simulation, mean square error criterion was used as a measure of deviation of the reconstructed image from the original one. On the bar graph of Fig. 5.10, d), one can also evaluate quantitatively the reconstruction accuracy and the speed of convergence of iterations. The graphs show that iterations converge quite rapidly, especially for coarser quantization and that the reconstruction accuracy for 255 phase quantization levels can be practically the same as that for kinoform without phase quantization. Coarse quantization of the kinoform phase may worsen the image intensity reconstruction quality substantially.

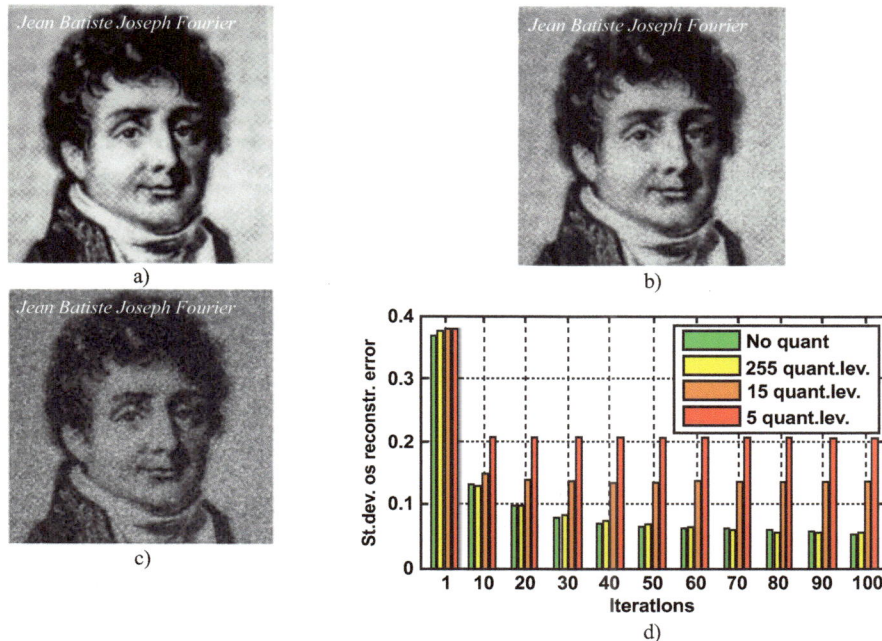

**Fig. 5.10.** Examples of images reconstructed from kinoforms: a) original image used for the synthesis of kinoforms; b) and c) - reconstructed images for kinoforms with 255 and 5 phase quantization levels, respectively; d) plots of standard deviation of the reconstruction error as a function of the number of iterations for no quantization, 255, 15 and 5 phase quantization levels (error values are normalized to the dynamic range [0,1]).

## 5.3   Encoding methods and introducing spatial carrier

All hologram encoding methods can be treated in a unified way as methods that explicitly or implicitly introduce to recorded holograms some form of a spatial carrier similarly to how optical holograms are recorded. This can be shown by writing Eqs.5.1.4, 5.1.7, 5.1.8 and 5.2.3 in the following equivalent form explicitly containing hologram samples multiplied by those of the spatial carrier with respect to one or both coordinates:

$$\tilde{\Gamma}_{m,n} = \tilde{\Gamma}_{2r+m_0,s} = \mathrm{Re}\left\{\Gamma_{r,s}\exp\left(-i2\pi\frac{m}{2}\right)\right\}\bigg/4 + b =$$

$$\tilde{\Gamma}_{2r+m_0,s} = \mathrm{Re}\left\{\Gamma_{r,s}\exp\left(-i2\pi\frac{2r+m_0}{2}\right)\right\}\bigg/4 + b\,;\ \ m = 2r+m_0;\ m_0 = 0,1;\ n = s\,;$$

$$(5.1.4')$$

$$\tilde{\Gamma}_{m,n} = \frac{1}{2}\mathrm{rctf}\left\{\mathrm{Re}\left[\Gamma_{r,s}\exp\left(-i2\pi\frac{m}{2}\right)\right]\right\} =$$

$$\tilde{\Gamma}_{4r+2m_{01}+m_{02},s} = \frac{1}{2}\mathrm{rctf}\left\{\mathrm{Re}\left[\Gamma_{r,s}\exp\left(-i2\pi\frac{4r+2m_{01}+m_{02}}{2}\right)\right]\right\};$$

$$m = 4r+2m_{01}+m_{02};\ m_{01},m_{02} = 0,1;\ n = s\,;\qquad\qquad (5.1.7')$$

$$\tilde{\Gamma}_{m,n} = \frac{1}{2}\mathrm{rctf}\left\{\mathrm{Re}\left[\Gamma_{r,s}\exp\left(-i2\pi\left(m+2n\right)/4\right)\right]\right\} =$$

$$\tilde{\Gamma}_{2r+m_0,2s+n_0} = \frac{1}{2}\mathrm{rctf}\left\{\mathrm{Re}\left[\Gamma_{r,s}\exp\left(-i2\pi\frac{2r+m_0}{4}\right)\exp\left(-i2\pi\frac{4s+n_0}{4}\right)\right]\right\}$$

$$m = 2r+m_0;\ n = 2s+n_0, m_0, n_0 = 0,1;\qquad\qquad (5.1.8')$$

$$\tilde{\Gamma}_{m,n} = A_0\exp\left\{i\left[\phi_{r,s} - \cos\left(2\pi\frac{m}{2}\right)\arccos\left(\left|\Gamma_{r,s}\right|/2A_0\right)\right]\right\} =$$

$$\tilde{\Gamma}_{2r+m_0,s} = A_0\exp\left\{i\left[\varphi_{r,s} - \cos\left(2\pi\frac{2r+m_0}{2}\right)\arccos\left(\left|\Gamma_{r,s}\right|/2A_0\right)\right]\right\}$$

$$m = 2r+m_0; m_0 = 0,1;\ n = s\,,\qquad\qquad (5.2.3')$$

where **rctf**$(X)$ is the "rectifier" function

$$\mathrm{rctf}\left(X\right) = \left\{\begin{array}{l} X,\ X \geq 0 \\ 0,\ X < 0 \end{array}\right.,\qquad\qquad (5.3.1)$$

As one can see from these expressions, the spatial carrier has a period no greater than one half that of hologram sampling. At least two samples of the spatial carrier have to correspond to one hologram sample in order to enable reconstruction of amplitude and phase of each hologram sample through the modulated signal of the spatial carrier. This redundancy implies that, in order to modulate the spatial carrier by a hologram, one or more intermediate samples are required between the basic ones. They must be determined by hologram interpolation.

The simplest interpolation method, the zero order, or nearest neighbor, interpolation, simply repeats samples. For instance, according to Eq.(5.1.4') each hologram sample is repeated twice for two samples of the spatial carrier, in Eq.(5.1.7') it is repeated four times for four samples, etc. Of course such an interpolation found in all modifications of the detour phase method yields a very rough approximation of hologram intermediate samples. In Ch. 7 it will be shown that the distortions of the reconstructed image caused by non-perfect interpolation of hologram samples manifest themselves in aliasing effects.

Aliasing free interpolation can be achieved with discrete sinc-interpolation ([6]). Discrete sinc-interpolated values of the desired intermediate samples can be obtained by using, at the hologram synthesis, Shifted DFT with appropriately found shift parameters that correspond to the position of required intermediate samples of the hologram. It should be also noted that the symmetrization method may be regarded as an analog of the method of Eq.(5.1.4) with discrete sinc-interpolation of intermediate samples. In the symmetrization method, such an interpolation is secured automatically and the restored images are free of aliasing images as it will be shown in Chapt. 7, Sect. 7.3.

Along with methods that introduce spatial carriers implicitly, explicit introduction of spatial carriers is also possible that directly simulates optical recording of holograms and interferograms. Of the methods oriented to amplitude-only media, one can mention those of Burch ([11]):

$$\tilde{\Gamma}_{m,n} = 1 + \left|\Gamma_{m,n}\right| \cos\left(2\pi\, fm + \phi_{m,n}\right) \tag{5.3.2}$$

and of Huang and Prasada ([12])

$$\tilde{\Gamma}_{m,n} = \left|\Gamma_{m,n}\right| \left[1 + \cos\left(2\pi\, fm + \phi_{m,n}\right)\right] \tag{5.3.3}$$

where $f$ is frequency of the spatial carrier.

Of binary-media-oriented methods, one can mention the method described in the overview [13]:

$$\tilde{\Gamma}_{m,n} = \text{hlim}\left\{\cos\left[\arcsin\left(\left|\Gamma_{m,n}\right|/A_0\right)\right] + \cos\left(2\pi\, fm + \varphi_{m,n}\right)\right\}, \tag{5.3.4}$$

where $A_0$ is maximal value of $\left|\Gamma_{m,n}\right|$.

Of the phase-media-oriented methods, we can mention that of Kirk and Jones ([14]):

$$\tilde{\Gamma}_{m,n} = A_o \exp\left\{i\left[\varphi_{m,n} - h_{m,n}\cos\left(2\pi\, fm\right)\right]\right\} \tag{5.3.5}$$

where $h_{m,n}$ depends in a certain way on $\left|\Gamma_{m,n}\right|$ and on the diffraction order where the reconstructed image should be obtained. In a sense, this method is equivalent to the multi phase encoding method. When $f = 1/2$ and

$$h_{m,n} = \arccos\left(\left|\Gamma_{m,n}\right|/2A_0\right) \tag{5.3.7}$$

it coincides with the double phase encoding method according to Eq. 5.2.3.

## 5.4   Artificial diffusers in the synthesis of display holograms

One problem common to all hologram recording methods is extremely high dynamic range (the ratio of maximal to minimal non-zero component) of mathematical Fourier holograms of regular images because intensity of image low frequency components is several order or magnitude larger than that of middle frequencies and this discrepancy grows for high frequencies. This phenomenon is illustrated in Fig. 5.11, b). At the same time, the dynamic range of best spatial light modulators is only of the order $10^2 \div 10^3$. Of the same order in also the number of quantization levels in digital-to-analog converters used to control spatial light modulators. One can fit the dynamic range of the hologram and hologram recording device by a non-linear compressing signal transformation such as, for instance, **P**-th law nonlinearity ([3]).

$$OUTPUT = \left[ abs\left(INPUT\right)\right]^P sign\left(INPUT\right),$$
(5.4.1)

where $P \le 1$ is a dynamic range compressing parameter, $abs(.)$ and $sign(.)$ are absolute value and sign of the variable. However the compressive nonlinear transformation will redistribute signal energy in favor of its high frequency components, which will result in restoration of only image contours as one can see on Fig. 5.8, d).

**Fig. 5.11.** Dynamic range of Fourier spectra of images and synthesis of hologram with artificial diffuser: a) – test image; b) – module of Fourier spectrum of the test image raised to a power 0.1 to enable its display in the compressed dynamic range; c) – module of Fourier spectrum of the same image with assigned to it a pseudo-random non-correlated phase component; d) image reconstructed from a Fourier hologram of the test image without pseudo-random phase component and recorded using the nonlinear compressive transform; e) – a result of reconstruction of a Fourier hologram of the test image with a pseudo-random phase component.

An alternative solution is assigning to images artificial phase component to make image spectrum almost uniform. In the synthesis of holograms, one has to specify the object wave front amplitude and phase. The amplitude component is defined by the image to be displayed. The object wave front phase component is irrelevant to visual observation, although it affects very substantially the object wave front spectrum dynamic range. Therefore one can select an object wave front phase distribution in such a way as to secure least possible distortions of the object's hologram due to the limitation of the hologram dynamic range and quantization in the process of hologram recording. The simplest way to do this is use arrays of pseudo-random non-correlated numbers uniformly distributed in the range $[-\pi,\pi]$ as a phase component, which makes image Fourier spectrum statistically homogeneous over all range of spatial frequencies (see an example in Fig. 5. 11, c)). Mathematical holograms of such images are less vulnerable to quantization in the process of recording, and recorded holograms reconstruct images with fewer distortions than holograms synthesized without assigning to images pseudo-random phase distribution as one can see comparing images d) and e).

Distortions of the hologram in the process of encoding and recording result in distortions of reconstructed images. For holograms synthesized with the use of the artificial phase component, or diffuser, image distortions exhibit themselves as speckle noise ([3]). In principle, these distortions can be minimized using above described iterative algorithm for synthesis of kinoform.

Assigning to the object wave front a pseudo-random phase component imitates diffuse properties of real object to scatter light. Non-correlated pseudo-random phase component corresponds to uniform scattering of light in all directions. One can also imitate different other types non-uniform scattering using pseudo-random phase components with correspondingly selected correlation function. We refer to such correlated pseudo-random phase component assigned to the object wave front with a purpose of imitating pre-assigned non-uniform light scattering as to ***programmed diffuser***. Using holograms with programmed diffuser for synthesis of display holograms of three-dimensional objects is discussed in Chapt. 9.

Algorithm for generating pseudo-random phase masks with pre-assigned correlation function is similar to the above-described algorithm of generating the optimized diffuser for synthesis of kinoform. Flow diagram of the algorithm is presented in Fig. 5. 12. In this diagram, $\{\vartheta_{k,l}\}$ is an array of numbers in the range $(-\pi,\pi)$ representing the required phase mask and $\{P(r,s)\}$ is an array of coefficients of representation of the required distribution of light intensity in the Fourier domain, or Discrete Fourier Transform of the required correlation function of the diffuser. By definition, arrays $\{\vartheta_{k,l}\}$ and $\{P(r,s)\}$ should satisfy the equation:

$$P(r,s) = C \cdot \left| \text{DFT}\left(\left\{\exp\left(i\vartheta_{k,l}\right)\right\}\right)\right|^2 \qquad (5.4.2)$$

where $C$ is an appropriate normalizing constant. Because, in general, the exact solution of this equation for $\{\vartheta_{k,l}\}$ given $\{P(r,s)\}$ may not exist, its approximation is sought by means of iterative minimization of an appropriate measure $D(\cdot,\cdot)$ of deviation of $C\cdot\left|\mathrm{DFT}\left\{\exp\left(i2\pi\hat{\vartheta}_{k,l}\right)\right\}\right|^{2}$ from $\left\{P(r,s)\right\}$:

$$\left\{\hat{\vartheta}_{k,l}\right\}=\arg\min_{\left\{\vartheta_{k,l}\right\}}\left[D\left(\left\{P(r,s)\right\},C\cdot\left|\mathrm{DFT}\left[\left\{\exp\left(i2\pi\hat{\vartheta}_{k,l}\right)\right\}\right]\right|^{2}\right)\right]. \tag{5.4.3}$$

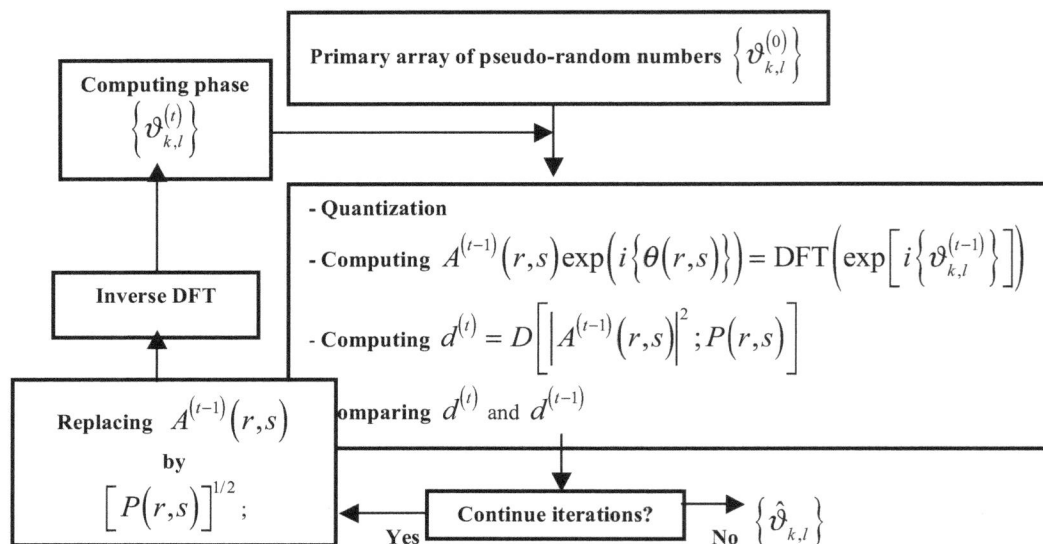

**Fig. 5.12.** Flow diagram of an iterative algorithm for generating pseudo-random phase masks with a specified power spectrum, or discrete Fourier transform of its autocorrelation function

## References

1. B. R. Brown, A. Lohmann, Complex spatial filtering with binary masks, Appl. Optics, 5, No. 6, 967-969 (1966)
2. B. R. Brown, A. Lohmann, Computer generated binary holograms, IMB J. Res. Dev., v. 13, No. 2, 160-168 (1969)
3. L. Yaroslavsky, Digital Holography and Digital Image Processing, Kluwer Academic Publishers, Boston, 2004
4. W. I. Dallas, Computer generated holograms, in: The Computer in Optical Research, Methods and Applications, B. R. Friede, Ed., Topics in Applied Physics, v. 41, Springer, Berlin, 1980, pp. 297-367
5. C. B. Burckhardt, A simplification of Lee's method of generating holograms by computer, Appl. Opt., v. 9, No.8, p. 1949 (1970)
6. L. Yaroslavskii, N. Merzlyakov, Methods of Digital Holography, Consultance Bureau, N.Y., 1980
7. C. K. Hsueh, A. A. Sawchuk, Computer generated double phase holograms,, Appl. Opt., v. 17, No. 24, pp. 3874-3883, 1978
8. L. B. Lesem, P. M. Hirsch, J. A. Jordan, Kinoform, IBM Journ. Res. Dev., 13, 150, 1969
9. N. C. Gallagher, B. Liu, Method for Computing Kinoforms that Reduces Image Reconstruction Error, Applied Optics, v. 12, No. 10, Oct. 1973, p. 2328
10. J. J. Burch, Algorithms for computer synthesis of spatial filters, Proc. IEEE, v. 55, 999, 1967
11. T. S. Huang, E. Prasada, Considerations on the generation and processing of holograms by digital computers, MIT/RLE Quat. Progr. Rept., 1966, v. 81
12. J. P. Kirk, A. L. Jones, Phase-only complex valued spatial filter, JOSA, v. 61, No. 8, pp. 1023-1028, 1971

*Digital Signal Processing in Experimental Research, 2009, 1, 89-109*

## 6.  Optical Reconstruction of Computer-Generated Holograms

**Abstract:** Computer-generated holograms are used as optical elements in optical setups. This chapter analyses how methods of generating, encoding and recording computer-generated hologram affect the results of their optical reconstruction.

At the reconstruction stage, computer generated holograms synthesized with the use of discrete representations of wave propagation transformations described in Ch. 2 and recorded using one of the hologram encoding methods described in Ch. 5 are subjected to analog optical transformations in reconstruction optical setups. The discrepancy between optical transforms and their discrete representation affects in a certain way the hologram reconstruction result. Encoding of mathematical holograms into physical holograms according to the method of hologram recording has its influence as well. In this chapter we discuss these issues. Specifically, we consider reconstruction, in analog optical set-ups performing optical Fourier transform shown in Fig. 6.1 of computer generated Fourier holograms for four methods of hologram encoding: for the symmetrization method, for the orthogonal encoding method, for the double phase recording on a phase medium and for kinoform.

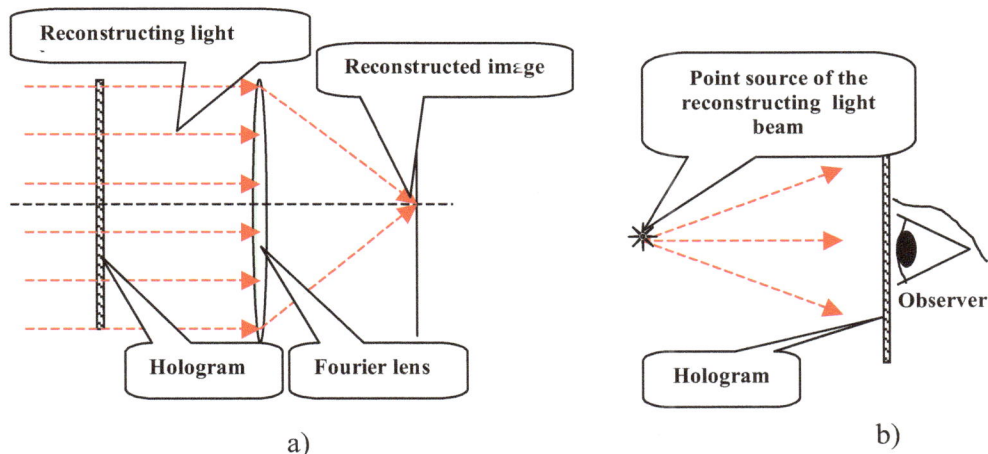

**Fig. 6.1.** Schemes of optical (a) and visual (b) reconstruction of computer generated Fourier holograms

### 6.1   Definitions and denotations

The following characteristics of the hologram recording device affect the reconstruction result: type of the sampling raster, sampling intervals, recording aperture and physical size of recorded holograms (see Fig. 6.2). In what follows, we will use the following denotations for them:

$(\xi, \eta)$      - physical coordinates on the hologram recording medium;

$(\Delta\xi, \Delta\eta)$  - sampling intervals of the rectangular sampling raster along coordinates $(\xi, \eta)$;

$\left(\xi_0, \eta_0\right)$  -shift parameters that depend on the geometry of positioning the hologram in the reconstruction set up;

$h_{rec}\left(\xi, \eta\right)$ -hologram recording device aperture function;

$w\left(\xi, \eta\right)$  - a window function that defines physical size of the recorded hologram: $0 < w\left(\xi, \eta\right) \leq 1$, when $\left(\xi, \eta\right)$  belong to the hologram area and $w\left(\xi, \eta\right) = 0$, otherwise.

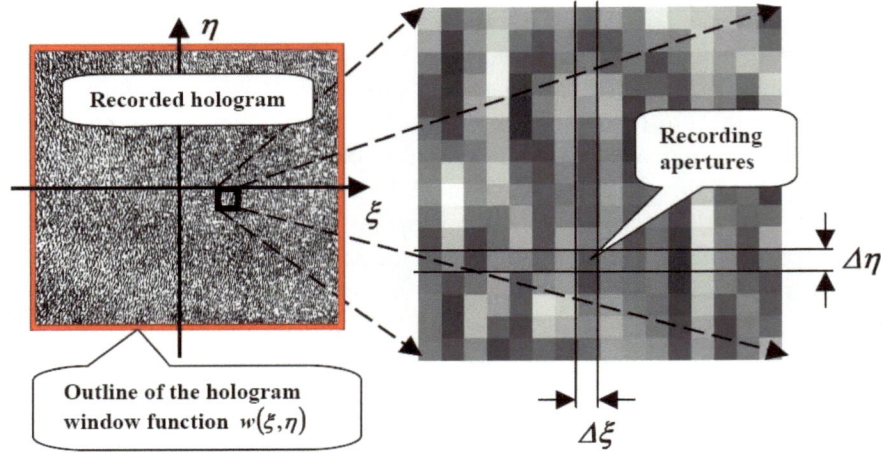

**Fig. 6.2.** Definitions related to recorded physical computer generated holograms

## 6.2    Reconstruction of Fourier holograms synthesized using the symmetrization method

For the symmetrization method, the following equation describes conversion of the numerical matrix of the mathematical hologram $\left\{ \Gamma_{r,s} \right\}$ into the physical hologram $\tilde{\Gamma}\left(\xi, \eta\right)$ recorded on an amplitude-only medium:

$$\tilde{\Gamma}\left(\xi, \eta\right) = w\left(\xi, \eta\right) \sum_r \sum_s \left(\Gamma_{r,s} + b\right) h_{rec}\left(\xi - \xi_0 - r\Delta\xi, \eta - \eta_0 - s\Delta\eta\right), \qquad (6.2.1)$$

where $b$ is a constant bias required for eliminating negative values in recording samples $\left\{ \Gamma_{r,s} \right\}$ of the mathematical hologram and indices $\left(r,s\right)$ run over all available hologram samples.  Eq. 6.2.1 can be represented as a in a convolution:

$$\tilde{\Gamma}\left(\xi, \eta\right) = w\left(\xi, \eta\right) \cdot \left\{ h_{rec}\left(\xi, \eta\right) \otimes \sum_{r=-\infty}^{\infty} \sum_{s=\infty}^{\infty} \left(\Gamma_{r,s} + b\right) \delta\left(\xi - \xi_0 - r\Delta\xi\right) \delta\left(\eta - \eta_0 - s\Delta\eta\right) \right\},$$

$$(6.2.2)$$

where $\otimes$ stands for the convolution and summation limits are changed to $\left[-\infty,\infty\right]$ bearing in mind that hologram window function $w(\xi,\eta)$ selects available hologram samples from virtual infinite number of samples.

Let now $(x,y)$ be coordinates in reconstructed image plane situated at a distance $Z$ from the hologram plane and $\lambda$ be wavelength of the reconstructing light beam. As it was mentioned, we assume that recorded computer generated holograms $\tilde{\Gamma}(\xi,\eta)$ is subjected at reconstruction to optical integral Fourier Transform. This results in reconstructed wave front $A_{rcstr}(x,y)$

$$A_{rcstr}(x,y) = \int\limits_{-\infty}^{\infty}\int\limits_{-\infty}^{\infty} \tilde{\Gamma}(\xi,\eta)\exp\left(-i2\pi\frac{x\xi+y\eta}{\lambda Z}\right)d\xi\,d\eta \qquad (6.2.3)$$

that can be represented, according to the convolution theorem for the integral Fourier transform, as:

$$A_{rcstr}(x,y) = W(x,y)\otimes\left\{\int\limits_{-\infty}^{\infty}\int\limits_{-\infty}^{\infty} h_{rec}(\xi,\eta)\exp\left(-i2\pi\frac{x\xi+y\eta}{\lambda Z}\right)d\xi\,d\eta\times\right.$$

$$\int\limits_{-\infty}^{\infty}\int\limits_{-\infty}^{\infty}\sum_{r=-\infty}^{\infty}\sum_{s=-\infty}^{\infty}(\Gamma_{r,s}+b)\delta(\xi-\xi_0-r\Delta\xi)\delta(\eta-\eta_0-s\Delta\eta)\exp\left(-i2\pi\frac{x\xi+y\eta}{\lambda Z}\right)d\xi\,d\eta\Bigg\} =$$

$$W(x,y)\otimes\left\{\int\limits_{-\infty}^{\infty}\int\limits_{-\infty}^{\infty} h_{rec}(\xi,\eta)\exp\left(-i2\pi\frac{x\xi+y\eta}{\lambda Z}\right)d\xi\,d\eta\times\right.$$

$$\sum_{r=-\infty}^{\infty}\sum_{s=-\infty}^{\infty}(\Gamma_{r,s}+b)\int\limits_{-\infty}^{\infty}\int\limits_{-\infty}^{\infty}\exp\left(-i2\pi\frac{x\xi+y\eta}{\lambda Z}\right)\delta(\xi-\xi_0-r\Delta\xi)\delta(\eta-\eta_0-s\Delta\eta)d\xi\,d\eta\Bigg\} =$$

$$W(x,y)\otimes\left\{H_{rec}(x,y)\sum_{r=-\infty}^{\infty}\sum_{s=-\infty}^{\infty}(\Gamma_{r,s}+b)\exp\left(-i2\pi\frac{r\Delta\xi x+s\Delta\eta y}{\lambda Z}\right)\right) \qquad (6.2.4)$$

where

$$W(x,y) = \int\limits_{-\infty}^{\infty}\int\limits_{-\infty}^{\infty} w(\xi,\eta)\exp\left(-i2\pi\frac{x\xi+y\eta}{\lambda Z}\right)d\xi\,d\eta \qquad (6.2.5)$$

and

$$H_{rec}(x,y) = \int\limits_{-\infty}^{\infty}\int\limits_{-\infty}^{\infty} h_{rec}(\xi,\eta)\exp\left[-i2\pi\frac{x(\xi-\xi_0)+y(\eta-\eta_0)}{\lambda Z}\right]d\xi\,d\eta \qquad (6.2.6)$$

are Fourier transforms of the hologram window function and of the aperture function of the hologram recording device, the latter being found with an account of the shifts

$(\xi_0, \eta_0)$ in positions of recorded hologram samples with respect to optical axis of the optical set up used for reconstruction of the hologram.

Introduce now an array $\overline{A}_{k,l}^{(o)}$, $k = 0,...,2N_1 - 1$, $l = 0,1,...,N_2 - 1$, of samples of the object wave front. In the symmetrization method, the array is defined by Eq. 5.1.13 (Chapt.6):

$$\overline{A}_{k,l}^{(o)} = \begin{cases} \overline{A}_{k,l}, 0 \leq k \leq N_1 - 1; 0 \leq l \leq N_2 - 1 \\ \overline{A}_{2N-k,l}, N_1 \leq k \leq 2N_1 - 1; 0 \leq l \leq N_2 - 1 \end{cases} \tag{6.2.7}$$

Let, for Fourier holograms, mathematical hologram $\Gamma_{r,s}$ be computed as Shifted DFT of this array:

$$\Gamma_{r,s} \propto \sum_{k=0}^{2N_1-1} \sum_{l=0}^{N_2-1} \overline{A}_{k,l}^{(o)} \exp\left\{ i2\pi\left[ \frac{(k+u)(r+p)}{2N_1} + \frac{(l+v)(s+q)}{N_2} \right] \right\}, \tag{6.2.8}$$

where $(u,v)$ and $(p,q)$ are transform shift parameters. Substitute Eq. 6.2.8 into Eq. 6.2.4 and obtain:

$$A_{rcstr}(x,y) \propto W(x,y) \otimes \{ H_{rec}(x,y) \times$$

$$\left\{ \sum_{r=-\infty}^{\infty} \sum_{s=-\infty}^{\infty} \left\{ \sum_{k=0}^{2N_1-1} \sum_{l=0}^{N_2-1} \overline{A}_{k,l}^{(o)} \exp\left\{ i2\pi\left[ \frac{(k+u)(r+p)}{2N_1} + \frac{(l+v)(s+q)}{N_2} \right] \right\} + b \right\} \times$$

$$\exp\left( -i2\pi \frac{r\Delta\xi x + s\Delta\eta y}{\lambda Z} \right) \right\} = W(x,y) \otimes \{ H_{rec}(x,y) \times$$

$$\left\{ \sum_{k=0}^{2N_1-1} \sum_{l=0}^{N_2-1} \overline{A}_{k,l}^{(o)} \exp\left[ i2\pi\left( \frac{k+u}{2N_1}p + \frac{l+v}{N_2}q \right) \right] \sum_{r=-\infty}^{\infty} \exp\left[ i2\pi\left( \frac{k+u}{2N_1} - \frac{\Delta\xi x}{\lambda Z} \right)r \right] \times$$

$$\sum_{s=-\infty}^{\infty} \exp\left[ i2\pi\left( \frac{l+v}{2N_1} - \frac{\Delta\eta y}{\lambda Z} \right)s \right] + b \sum_{r=-\infty}^{\infty} \exp\left( -i2\pi \frac{\Delta\xi x}{\lambda Z}r \right) \sum_{s=-\infty}^{\infty} \exp\left( -i2\pi \frac{\Delta\eta y}{\lambda Z}s \right) \}.$$
(6.2.9)

At this stage one can see that selection of shift parameters $p = q = 0$ will be appropriate to remove phase shift factors at samples $\left\{ \overline{A}_{k,l}^{(o)} \right\}$ of the object wave front. Then, using the Poisson's summation formula,:

$$\sum_{r=-\infty}^{\infty} \exp(-i2\pi fr) = \sum_{do=-\infty}^{\infty} \delta(f - do) \tag{6.2.10}$$

obtain:

$$A_{rcstr}\left(x,y\right) \propto W\left(x,y\right) \otimes$$

$$\left\{ H_{rec}\left(x,y\right) \left\{ \sum_{do_x=-\infty}^{\infty} \sum_{do_y=-\infty}^{\infty} \sum_{k=0}^{2N_1-1} \sum_{l=0}^{N_2-1} \overline{A}_{k,l}^{(o)} \delta\left(\frac{k+u}{2N_1} - \frac{\Delta\xi x}{\lambda Z} - do_x\right)\delta\left(\frac{l+v}{N_2} - \frac{\Delta\eta y}{\lambda Z} - do_y\right) + \right.\right.$$

$$\left.\left. b \sum_{do_x=-\infty}^{\infty} \sum_{do_y=-\infty}^{\infty} \delta\left(\frac{\Delta\xi x}{\lambda Z} - do_x\right)\delta\left(\frac{\Delta\eta y}{\lambda Z} - do_y\right)\right\}\right\} =$$

$$\sum_{do_x=-\infty}^{\infty} \sum_{do_y=-\infty}^{\infty} \tilde{\overline{A}}^{(o)}\left(x,do_x;y,do_y\right) +$$

$$b \sum_{do_x=-\infty}^{\infty} \sum_{do_y=-\infty}^{\infty} W\left(x - \frac{\lambda Z}{\Delta\xi}do_x, y - \frac{\lambda Z}{\Delta\xi}do_y\right)\tilde{H}_{rec}^{(\xi,\eta_0)}\left(\frac{\lambda Z}{\Delta\xi}do_x, \frac{\lambda Z}{\Delta\xi}do_y\right), \quad (6.2.11)$$

where it is denoted:

$$\tilde{\overline{A}}^{(o)}\left(x,do_x;y,do_y\right) = \sum_{k=0}^{2N_1-1} \sum_{l=0}^{N_2-1} \tilde{H}_{rec}\left[\left(\frac{k+u}{N_1} - do_x\right)\frac{\lambda Z}{\Delta\xi}, \left(\frac{l+v}{N_2} - do_y\right)\frac{\lambda Z}{\Delta\xi}\right]\overline{A}_{k,l}^{(o)} \times$$

$$W\left(x - \frac{k+u}{N_1}\frac{\lambda Z}{\Delta\xi} - do_x\frac{\lambda Z}{\Delta\xi}; x - \frac{l+v}{N_2}\frac{\lambda Z}{\Delta\xi} - do_y\frac{\lambda Z}{\Delta\xi}\right), \quad (6.2.12)$$

More detailed derivation of Eq. 6.2.11 is provided in Appendix.

It follows from this formula that
- hologram placed into the optical Fourier system reconstructs the original object wave front in several diffraction orders defined by indices $do_x$ and $do_y$;
- pattern of the reconstructed images samples $\tilde{\overline{A}}^{(o)}\left(x,do_x;y,do_y\right)$ in the diffraction orders is masked by a function $\tilde{H}_{rec}^{(\xi_0,\eta_0)}\left(\frac{\lambda Z}{\Delta\xi}do_x, \frac{\lambda Z}{\Delta\xi}do_y\right)$, which is the frequency response of the hologram recording device.
- the masked object wave front is reconstructed by interpolation of its samples with the interpolation function $W\left(x - \frac{\lambda Z}{\Delta\xi}do_x, y - \frac{\lambda Z}{\Delta\xi}do_y\right)$ equal to Fourier transform of the hologram window function $w(\xi,\eta)$;
- constant bias in hologram recording results in appearance in the reconstructed image of bright spots in the center of each diffraction order (the last term in Eq. 6.2.11)

Arrangement of diffraction orders in the reconstructed image plane under shift parameters $u = -N_1, v = -N_2/2$ and an example of an image optically reconstructed from a computer-generated hologram synthesized using object symmetrization by duplication are shown in Fig. 6.3.

Note that Eq. 6.2.10 can be extended to the case of hologram recording on combined amplitude and phase medium. For this one can simply discard in the formula the term containing constant bias $b$:

$$A_{rcstr}(x,y) = \sum_{k=0}^{2N_1-1} \sum_{l=0}^{N_2-1} \sum_{do_x=-\infty}^{\infty} \sum_{do_y=-\infty}^{\infty} \tilde{H}_{rec}\left[\left(\frac{k+u}{N_1} - do_x\right)\frac{\lambda Z}{\Delta\xi}, \left(\frac{l+v}{N_2} - do_y\right)\frac{\lambda Z}{\Delta\xi}\right] A_{k,l}^{(o)} \times$$

$$W\left(x - \frac{k+u}{N_1}\frac{\lambda Z}{\Delta\xi} - do_x\frac{\lambda Z}{\Delta\xi}; x - \frac{l+v}{N_2}\frac{\lambda Z}{\Delta\xi} - do_y\frac{\lambda Z}{\Delta\xi}\right) = \sum_{do_x=-\infty}^{\infty} \sum_{do_y=-\infty}^{\infty} \tilde{\tilde{A}}^{(o)}\left(x, do_x; y, do_y\right)$$

$$\tag{6.2.13}$$

where $\left\{A_{k,l}^{(o)}\right\}$ are samples of the object wave front complex amplitude.

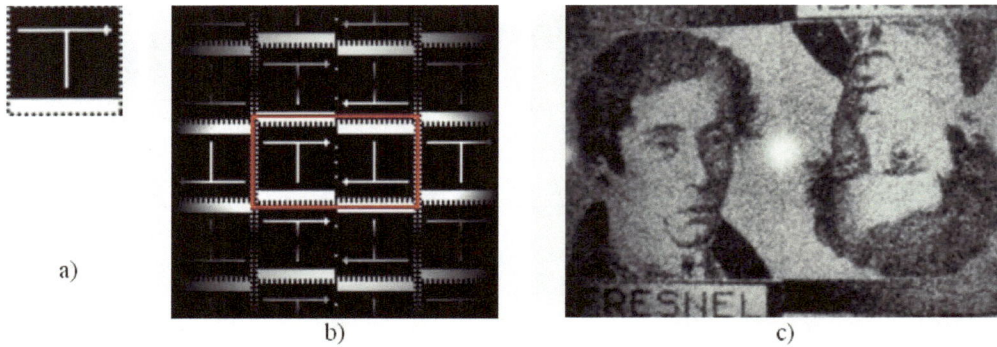

**Fig. 6.3.** Reconstruction of a Fourier hologram synthesized using image symmetrization by duplication: a) - a test object, b) - diffraction orders of reconstruction masked by the frequency response of the hologram recording device; the zero order is outlined by the red rectangle; c) – an example of optical reconstruction of a computer generated hologram synthesized using symmetrization by duplication; bright spot in the center results from the constant bias in hologram recording.

## 6.3   Reconstruction of holograms recorder using the orthogonal encoding method

In case of orthogonal encoding, the encoded hologram $\tilde{\Gamma}_{m,n}$ is formed from samples $\Gamma_{r,s}$ of the mathematical hologram as

$$\tilde{\Gamma}_{m,n} = \frac{1}{2}(-1)^r (i)^{m_0}\left[(-1)^{m_0}\Gamma_{r,s} + \Gamma_{r,s}^*\right] + b \tag{6.3.1}$$

where $m = 2r + m_0$, $m_0 = 0,1$, $n = r$ (Chapt. 5, Eq. 5.1.4). Then for the physical hologram $\tilde{\Gamma}(\xi,\eta)$ in continuous coordinates $(\xi,\eta)$ of the recording medium one obtains, in the same denotations that were adopted above:

$$\tilde{\Gamma}(\xi,\eta) = \frac{1}{2}w(\xi,\eta)\left\{\sum_{r=-\infty}^{\infty} \sum_{s=-\infty}^{\infty} \sum_{m_0=0}^{1}\left\{(-1)^r (i)^{m_0}\left[(-1)^{m_0}\Gamma_{r,s} + \Gamma_{r,s}^*\right] + b\right\}\right\} \times$$

$$h_{rec}\left(\xi-\xi_0-r\Delta\xi,\eta-\eta_0-s\Delta\eta\right)\Big\}=$$

$$w(\xi,\eta)\Bigg\{\sum_{r=-\infty}^{\infty}\sum_{s=-\infty}^{\infty}\sum_{m_0=0}^{1}\Big\{\mathrm{Re}\Big[(-1)^r(-i)^{m_0}\,\Gamma_{r,s}\Big]+b\Big\}\,h_{rec}\left(\xi+\xi_0-r\Delta\xi,\eta+\eta_0-s\Delta\eta\right)\Bigg\}=$$

$$w(\xi,\eta)\Bigg\{h_{rec}(\xi,\eta)\otimes\sum_{r=-\infty}^{\infty}\sum_{s=-\infty}^{\infty}\sum_{m_0=0}^{1}\Big\{\mathrm{Re}\Big[(-1)^r(-i)^{m_0}\,\Gamma_{r,s}\Big]+b\Big\}\times$$

$$\delta\left(\xi-\xi_0-r\Delta\xi\right)\delta\left(,\eta-\eta_0-s\Delta\eta\right)\Bigg\}, \qquad (6.3.2)$$

where $\mathrm{Re}(.)$ is real part of the variable.

Following the reasoning and settings used in deriving Eq.(6.2.10), one can obtain that, in the Fourier transform reconstruction scheme, such a hologram reconstructs the following wave front:

$$A_{rcstr}(x,y)\propto W(x,y)\otimes\Big\{H_{rec}(x,y)\times$$

$$\Bigg\{\cos\left[\pi\left(\frac{1}{4}+\frac{\Delta\xi x}{\lambda Z}\right)\right]\sum_{m=-\infty}^{\infty}\sum_{n=-\infty}^{\infty}\sum_{k=0}^{N_1}\sum_{l=0}^{N_2}\overline{A}_{k,l}^{(o)}\,\delta\left(\frac{k-u}{N_1}-\frac{\Delta\xi x}{\lambda Z}-do_x+\frac{1}{2},\frac{l+v}{N_2}-\frac{\Delta\eta y}{\lambda Z}-do_y\right)+$$

$$\sin\left[\pi\left(\frac{1}{4}-\frac{\Delta\xi x}{\lambda Z}\right)\right]\sum_{m=-\infty}^{\infty}\sum_{n=-\infty}^{\infty}\sum_{k=0}^{N_1}\sum_{l=0}^{N_2}\overline{A}_{N_1-k,N_2-l}^{(o)}\,\delta\left(\frac{k+u}{N_1}-\frac{\Delta\xi x}{\lambda Z}-do_x+\frac{1}{2},\frac{l+v}{N_2}-\frac{\Delta\eta y}{\lambda Z}-do_y\right)+$$

$$b\cos\left(\pi\frac{\Delta\xi x}{\lambda Z}\right)\sum_{m=-\infty}^{\infty}\sum_{n=-\infty}^{\infty}\delta\left(\frac{\Delta\xi x}{\lambda Z}-do_x+\frac{1}{2},\frac{\Delta\eta y}{\lambda Z}-do_y\right)\Bigg\}\Bigg\}=$$

$$\cos\left[\pi\left(\frac{1}{4}+\frac{\Delta\xi x}{\lambda Z}\right)\right]\sum_{m=-\infty}^{\infty}\sum_{n=-\infty}^{\infty}\tilde{\tilde{A}}_{dir}^{(o)}\left(x,do_x;y,do_y\right)+$$

$$\sin\left[\pi\left(\frac{1}{4}-\frac{\Delta\xi x}{\lambda Z}\right)\right]\sum_{m=-\infty}^{\infty}\sum_{n=-\infty}^{\infty}\tilde{\tilde{A}}_{conj}^{(o)}\left(x,do_x;y,do_y\right)+$$

$$b\cos\left(\pi\frac{\Delta\xi x}{\lambda Z}\right)\sum_{m=-\infty}^{\infty}\sum_{n=-\infty}^{\infty}W\left(\frac{\Delta\xi x}{\lambda Z}-do_x+\frac{1}{2},\frac{\Delta\eta y}{\lambda Z}-do_y\right)\Bigg\}\Bigg\}, \quad (6.3.3)$$

where

$$\tilde{\tilde{A}}_{dir}^{(o)}\left(x,do_x;y,do_y\right)=\sum_{k=0}^{2N_1-1}\sum_{l=0}^{N_2-1}\tilde{H}_{rec}\left[\left(\frac{k+u}{N_1}-do_x\right)\frac{\lambda Z}{\Delta\xi},\left(\frac{l+v}{N_2}-do_y\right)\frac{\lambda Z}{\Delta\xi}\right]\overline{A}_{k,l}^{(o)}\times$$

$$W\left(x-\frac{k+u}{N_1}\frac{\lambda Z}{\Delta\xi}-do_x\frac{\lambda Z}{\Delta\xi};x-\frac{l+v}{N_2}\frac{\lambda Z}{\Delta\xi}-do_y\frac{\lambda Z}{\Delta\xi}\right), \qquad (6.3.4)$$

$$\tilde{\bar{A}}_{conj}^{(o)}\left(x,do_x;y,do_y\right)=\sum_{k=0}^{2N_1-1}\sum_{l=0}^{N_2-1}\tilde{H}_{rec}\left[\left(\frac{k+u}{N_1}-do_x\right)\frac{\lambda Z}{\Delta\xi},\left(\frac{l+v}{N_2}-do_y\right)\frac{\lambda Z}{\Delta\xi}\right]\bar{A}_{N_1-k,N_2-l}^{(o)}\times$$

$$W\left(x-\frac{k+u}{N_1}\frac{\lambda Z}{\Delta\xi}-do_x\frac{\lambda Z}{\Delta\xi};x-\frac{l+v}{N_2}\frac{\lambda Z}{\Delta\xi}-do_y\frac{\lambda Z}{\Delta\xi}\right),\qquad(6.3.5)$$

and, as above, $(u,v)$ are shift parameters of SDFT used in computing the mathematical hologram (as in Eq. 6.2.8 with $(p,q)$ set, as above, to zero) and $N_1$ and $N_2$ are dimensions of the array $\left\{\bar{A}_{k,l}^{(o)}\right\}$.

One can see from Eq. (6.3.3) that, similarly to the above case, the reconstructed image contains a number of diffraction orders, masked by the function $\tilde{H}_{rec}\left[\left(\frac{k+u}{N_1}-do_x\right)\frac{\lambda Z}{\Delta\xi},\left(\frac{l+v}{N_2}-do_y\right)\frac{\lambda Z}{\Delta\xi}\right]$, the frequency response of the hologram recording device, and a central spot in several diffraction orders due to the constant bias in the hologram. But here, in contrast to the above case, each diffraction order contains two superimposed images of the object, - a direct image and its conjugate rotated by $180^o$ with respect the former. Each of them is additionally masked by functions $\left\{\cos\left[\pi\left(\frac{1}{4}+\frac{\Delta\xi x}{\lambda D}\right)\right]\right\}$ and $\left\{\sin\left[\pi\left(\frac{1}{4}-\frac{\Delta\xi x}{\lambda D}\right)\right]\right\}$, respectively.

Therefore, in the central region of the direct image the conjugate image is suppressed, but at its periphery the aliasing conjugate image has an intensity comparable with that of the direct image. This is caused by the fact, that in the orthogonal encoding method additional intermediate samples hologram needed for representing the spatial carrier are obtained by the least accurate nearest neighbor interpolation of samples of the mathematical hologram.

A pattern of diffraction orders of the direct and conjugate images and their corresponding masking functions are shown in Fig. 6.4, b) for $u=-N_1/2$ and $v=-N_2/2$.

One can easily see there the direct image and its conjugate, an aliasing image whose contrast increases along the horizontal axis from the center to the periphery. Additionally, Fig. 6.4, c) represents an example of an image optically reconstructed from a computer generated hologram with orthogonal encoding, in which one can also clearly see the aliasing artifacts.

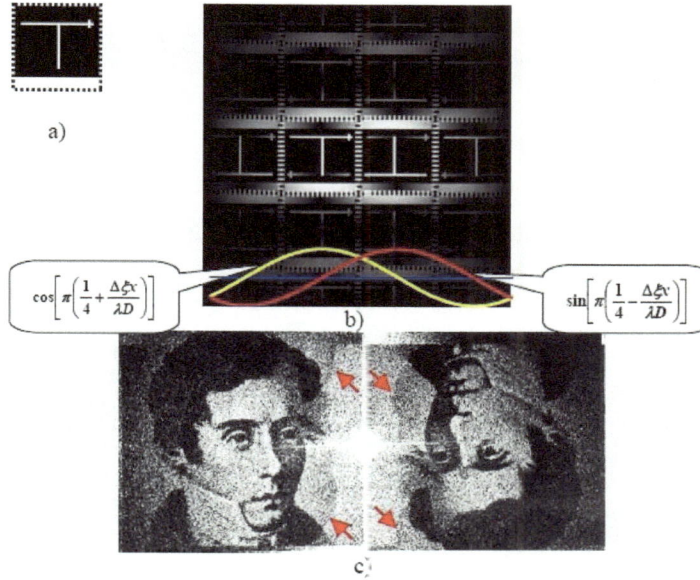

**Fig. 6.4.** Test image (a), its reconstruction (b) for the orthogonal encoding method (at the bottom, spatial weighting functions of the direct and conjugate images are depicted) and an example (c) of optical reconstruction of a computer generated hologram synthesized using the orthogonal coding method; red arrows indicate aliasing images superposed on the reconstructed image.

## 6.4    Reconstruction of holograms recorded on phase media using the double-phase method

The phase recording of the hologram on phase-only media according Eq. 5.2.3 (Chapt. 5) is described, in the same denotations, as

$$\tilde{\Gamma}(\xi,\eta) = w(\xi,\eta)\left\{ \sum_{r=-\infty}^{\infty} \sum_{s=-\infty}^{\infty} \sum_{m_o=0}^{1} \exp\left\{ i\left[ \phi_{r,s} + (-1)^{m_0} \arccos\frac{|\Gamma_{r,s}|}{2A_0} \right]\right\} \times \right.$$

$$h_{rec}\left[ \xi + \xi_0 - (2r + m_0)\Delta\xi, \eta + \eta_0 - s\Delta\eta \right] =$$

$$w(\xi,\eta)\left\{ h_{rec}(\xi,\eta) \otimes \sum_{r=-\infty}^{\infty} \sum_{s=-\infty}^{\infty} \sum_{m_o=0}^{1} \exp\left\{ i\left[ \phi_{r,s} + (-1)^{m_0} \arccos\left(\frac{|\Gamma_{r,s}|}{2A_0}\right) \right]\right\} \times \right.$$

$$\delta\left[ \xi - \xi_0 - (2r + m_0)\Delta\xi \right] \delta(\eta - \eta_0 - s\Delta\eta)\right\}. \qquad (6.4.1)$$

In a Fourier hologram reconstruction optical setup, this hologram is Fourier transformed and reconstructs the wave front described as

$$A_{rcstr}(x,y) = \int_{-\infty}^{\infty} \int_{-\infty}^{\infty} \tilde{\Gamma}(\xi,\eta)\exp\left( -i2\pi\frac{x\xi + y\eta}{\lambda Z} \right)d\xi\, d\eta =$$

$$W(x,y) \otimes \left\{ \int_{-\infty}^{\infty} \int_{-\infty}^{\infty} h_{rec}(\xi,\eta) \exp\left[ -i2\pi \frac{x(\xi-\xi_0) + y(\eta-\eta_0)}{\lambda Z} \right] d\xi\, d\eta \times \right.$$

$$\sum_{s=-\infty}^{\infty} \sum_{r=-\infty}^{\infty} \sum_{m_0=0}^{1} \exp\left\{ i\left[ \phi_{r,s} + (-1)^{m_0} \arccos\frac{|\Gamma_{r,s}|}{2A_0} \right] \right\} \times$$

$$\left. \int_{-\infty}^{\infty} \int_{-\infty}^{\infty} \exp\left( -i2\pi \frac{x\xi + y\eta y}{\lambda Z} \right) \delta\left[ \xi - \xi_0 - (2r+m_0)\Delta\xi \right] \delta(\eta - \eta_0 - s\Delta\eta)\, d\xi\, d\eta \right\} =$$

$$W(x,y) \otimes \left\{ \tilde{H}_{rec}(x,y) \sum_{s=-\infty}^{\infty} \sum_{r=-\infty}^{\infty} \sum_{m_0=0}^{1} \exp\left\{ i\left[ \phi_{r,s} + (-1)^{m_0} \arccos\frac{|\Gamma_{r,s}|}{2A_0} \right] \right\} \times \right.$$

$$\left. \exp\left[ i2\pi \frac{(2r+m_0)\Delta\xi x + s\Delta\eta y}{\lambda Z} \right] \right\} =$$

$$W(x,y) \otimes \left\{ \tilde{H}_{rec}(x,y) \sum_{s=-\infty}^{\infty} \sum_{r=-\infty}^{\infty} \exp\left\{ i\left[ \phi_{r,s} + (-1)^{m_0} \arccos\frac{|\Gamma_{r,s}|}{2A_0} \right] \right\} \times \right.$$

$$\left. \exp\left( i2\pi \frac{2r\Delta\xi x + s\Delta\eta y}{\lambda Z} \right) + \exp\left[ i2\pi \frac{(2r+1)\Delta\xi x + s\Delta\eta y}{\lambda Z} \right] \right\} =$$

$$W(x,y) \otimes \left\{ \tilde{H}_{rec}(x,y) \sum_{s=-\infty}^{\infty} \sum_{r=-\infty}^{\infty} \left\{ \exp\left[ i\left( \phi_{r,s} + \arccos\frac{|\Gamma_{r,s}|}{2A_0} \right) \right] \exp\left[ i2\pi \frac{2r\Delta\xi x + s\Delta\eta y}{\lambda Z} \right] + \right. \right.$$

$$\left. \left. + \exp\left[ i\left( \phi_{r,s} - \arccos\frac{|\Gamma_{r,s}|}{2A_0} \right) \right] \exp\left[ i2\pi \frac{(2r+1)\Delta\xi x + s\Delta\eta y}{\lambda Z} \right] \right\} \right\} =$$

$$W(x,y) \otimes \left\{ \tilde{H}_{rec}(x,y) \exp\left( i\pi \frac{\Delta\xi x}{\lambda Z} \right) \sum_{s=-\infty}^{\infty} \sum_{r=-\infty}^{\infty} \exp(i\varphi_{r,s}) \exp\left( i2\pi \frac{2r\Delta\xi x + s\Delta\eta y}{\lambda Z} \right) \times \right.$$

$$\left[ \exp\left( i\arccos\frac{|\Gamma_{r,s}|}{2A_0} \right) \exp\left( i\pi \frac{\Delta\xi x}{\lambda Z} \right) + \exp\left( -i\arccos\frac{|\Gamma_{r,s}|}{2A_0} \right) \exp\left( i\pi \frac{\Delta\xi x}{\lambda Z} \right) \right] =$$

$$2W(x,y) \otimes \left\{ \tilde{H}_{rec}(x,y) \exp\left( i\pi \frac{\Delta\xi x}{\lambda Z} \right) \sum_{s=-\infty}^{\infty} \sum_{r=-\infty}^{\infty} \exp(i\varphi_{r,s}) \exp\left( i2\pi \frac{2r\Delta\xi x + s\Delta\eta y}{\lambda Z} \right) \times \right.$$

$$\cos\left( \arccos\frac{|\Gamma_{r,s}|}{2A_0} + \pi\frac{\Delta\xi x}{\lambda Z} \right) =$$

$$2W\left(x,y\right)\otimes\left\{\tilde{H}_{rec}\left(x,y\right)\exp\left(i\pi\frac{\Delta\xi x}{\lambda Z}\right)\times\right.$$

$$\left[\cos\left(\pi\frac{\Delta\xi x}{\lambda Z}\right)\sum_{r=-\infty}^{\infty}\sum_{s=-\infty}^{\infty}\frac{\left|\Gamma_{r,s}\right|}{2A_0}\exp\left(i\varphi_{r,s}\right)\exp\left(i2\pi\frac{2r\Delta\xi x+s\Delta\eta y}{\lambda Z}\right)\right.$$

$$\left.\left.\sin\left(\pi\frac{\Delta\xi x}{\lambda Z}\right)\sum_{r=-\infty}^{\infty}\sum_{s=-\infty}^{\infty}\sqrt{1-\frac{\left|\Gamma_{r,s}\right|^2}{4A_0^2}}\exp\left(i\varphi_{r,s}\right)\exp\left(i2\pi\frac{2r\Delta\xi x+s\Delta\eta y}{\lambda Z}\right)\right]\right]. \quad (6.4.2)$$

By substituting $\Gamma_{r,s}$ with its expression through $SDFT\left(u,v;p,q\right)$:

$$\left|\Gamma_{r,s}\right|\exp\left(i\varphi_{r,s}\right)=\sum_{k=0}^{N_1-1}\sum_{l=0}^{N_2-1}\overline{A}_{k,l}^{(o)}\exp\left\{i2\pi\left[\frac{\left(k+u\right)\left(r+p\right)}{N_1}+\frac{\left(l+v\right)\left(s+q\right)}{N_2}\right]\right\}$$

$$(6.4.3)$$

and introducing an auxiliary function $\tilde{A}_{k,l}^{(0)}$ defined through equation

$$\sqrt{4A_0^2-\left|\Gamma_{r,s}\right|^2}\exp\left(i\varphi_{r,s}\right)=\sum_{k=0}^{N_1-1}\sum_{l=0}^{N_2-1}\tilde{A}_{k,l}^{(o)}\exp\left\{i2\pi\left[\frac{\left(k+u\right)\left(r+p\right)}{N_1}+\frac{\left(l+v\right)\left(s+q\right)}{N_2}\right]\right\}$$

$$(6.4.4)$$

obtain after some transformations and setting $p=q=0$:

$$A_{rcstr}\left(x,y\right)\propto W\left(x,y\right)\otimes\left\{H_{rec}\left(x,y\right)\exp\left(i\pi\frac{\Delta\xi x}{\lambda Z}\right)\times\right.$$

$$\left[\cos\left(\pi\frac{\Delta\xi x}{\lambda Z}\right)\sum_{m=-\infty}^{\infty}\sum_{n=-\infty}^{\infty}\sum_{k=0}^{N_1}\sum_{l=0}^{N_2}\overline{A}_{k,l}^{(o)}\delta\left(\frac{k+u}{N_1}-\frac{2\Delta\xi x}{\lambda Z}-do_x,\frac{l+v}{N_2}-\frac{\Delta\eta y}{\lambda Z}-do_y\right)+\right.$$

$$\left.\left.\sin\left(\pi\frac{\Delta\xi x}{\lambda Z}\right)\sum_{m=-\infty}^{\infty}\sum_{n=-\infty}^{\infty}\sum_{k=0}^{N_1}\sum_{l=0}^{N_2}\tilde{A}_{k,l}^{(o)}\delta\left(\frac{k+u}{N_1}-\frac{2\Delta\xi x}{\lambda Z}-do_x,\frac{l+v}{N_2}-\frac{\Delta\eta y}{\lambda Z}-do_y\right)\right]\right\}=$$

$$\cos\left(\pi\frac{\Delta\xi x}{\lambda Z}\right)\sum_{m=-\infty}^{\infty}\sum_{n=-\infty}^{\infty}\tilde{\overline{A}}^{(o)}\left(x,do_x;y,do_y\right)+\sin\left(\pi\frac{\Delta\xi x}{\lambda Z}\right)\sum_{m=-\infty}^{\infty}\sum_{n=-\infty}^{\infty}\tilde{\tilde{A}}^{(o)}\left(x,do_x;y,do_y\right),$$

$$(6.4.5)$$

where

$$\tilde{\overline{A}}^{(o)}\left(x,do_x;y,do_y\right)=\sum_{k=0}^{2N_1-1}\sum_{l=0}^{N_2-1}\tilde{H}_{rec}\left[\left(\frac{k+u}{N_1}-do_x\right)\frac{\lambda Z}{\Delta\xi},\left(\frac{l+v}{N_2}-do_y\right)\frac{\lambda Z}{\Delta\xi}\right]\overline{A}_{k,l}^{(o)}\times$$

$$W\left(x-\frac{k+u}{N_1}\frac{\lambda Z}{\Delta\xi}-do_x\frac{\lambda Z}{\Delta\xi};x-\frac{l+v}{N_2}\frac{\lambda Z}{\Delta\xi}-do_y\frac{\lambda Z}{\Delta\xi}\right), \quad (6.4.6)$$

$$\tilde{\bar{A}}^{(o)}\left(x, do_x; y, do_y\right) = \sum_{k=0}^{2N_1-1} \sum_{l=0}^{N_2-1} \tilde{H}_{rec}\left[\left(\frac{k+u}{N_1} - do_x\right)\frac{\lambda Z}{\Delta \xi}, \left(\frac{l+v}{N_2} - do_y\right)\frac{\lambda Z}{\Delta \xi}\right] \tilde{A}_{k,l}^{(o)} \times$$

$$W\left(x - \frac{k+u}{N_1}\frac{\lambda Z}{\Delta \xi} - do_x \frac{\lambda Z}{\Delta \xi}; x - \frac{l+v}{N_2}\frac{\lambda Z}{\Delta \xi} - do_y \frac{\lambda Z}{\Delta \xi}\right). \qquad (6.4.7)$$

The result of reconstruction of a hologram recorded on a phase medium by the two phase method is, thus, similar to the reconstruction of an hologram with orthogonal encoding (see Eq. 6.3.3). Image is also reconstructed in a number of diffraction orders masked by the function $H_{rec}(x, y)$. There is also superposition of the aliasing image described by the function $\tilde{A}_{k,l}^{(o)}$ over the original image $\bar{A}_{k,l}^{(o)}$ and the original and aliasing images are additionally masked by the functions $\cos(\pi \Delta \xi x / \lambda D)$ and $\sin(\pi \Delta \xi x / \lambda D)$, respectively. In the center of the proper image aliasing image is fully attenuated but over the peripheral area it may be of the same intensity as the proper image. In contrast to the orthogonal coding method, the aliasing image here is not conjugate to the original one, but is similar to it, in a sense, because, according to Eq. 6.4.4, it has the same phase spectrum and a distorted amplitude spectrum. An example of an aliasing image is shown in Fig. 6.5 for input images with and without adding pseudo-random phase components.

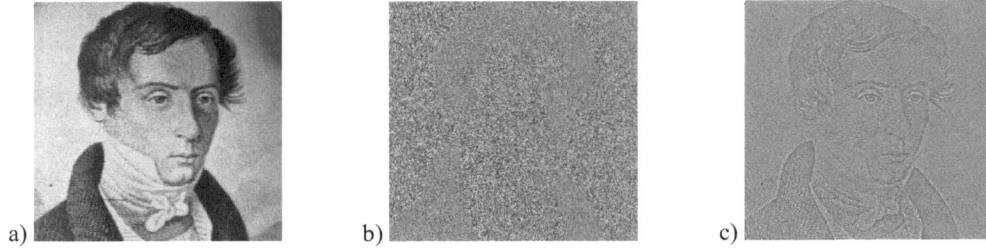

a)    b)    c)

**Fig. 6.5.** Input image (a) and aliasing images in double phase encoding method for the input image with and without pseudo-random phase component ( b) and c), correspondingly)

Unlike the holograms encoded by means of the symmetrization or orthogonal methods, double- phase encoding does not produce a central spot in the diffraction orders of the reconstructed image because holograms are recorded using this method on a phase medium without an amplitude bias.

## 6.5    Reconstruction of kinoforms

Let

$$\Gamma_{r,s} = \sum_{k=0}^{N_1-1} \sum_{l=0}^{N_2-1} \bar{A}_{k,l}^{(kf)} \exp\left\{i2\pi\left[\frac{(k+u)(r+p)}{N_1} + \frac{(l+v)(s+q)}{N_2}\right]\right\} \qquad (6.5.1)$$

be samples of the mathematical Fourier hologram generated through SDFT(*u,v;p,q*) from $N_1 \times N_2$ of object wave front samples $\left\{ \overline{A}_{k,l}^{(kf)} \right\}$ obtained from samples $\left\{ A_{k,l} \right\}$ of the object image in the process of kinoform encoding (these arrays of samples are exemplified by images in Fig. 5.10 a) and b), correspondingly). As kinoform samples contain only the phase component, they can be directly recorded onto a phase SLM, and the resulting recorded kinoform hologram can be written as

$$\tilde{\Gamma}(\xi,\eta) = w(\xi,\eta) \sum_r \sum_s \Gamma_{r,s} h_{rec}\left(\xi - \xi_0 - r\Delta\xi, \eta - \eta_0 - s\Delta\eta\right) \qquad (6.5.2)$$

At the reconstruction stage, kinoform is subjected to optical Fourier transform in an optical setup to reconstruct an image

$$A_{rcstr}(x,y) = \int_{-\infty}^{\infty} \int_{-\infty}^{\infty} \left[ w(\xi,\eta) \sum_r \sum_s \Gamma_{r,s} h_{rec}\left(\xi - \xi_0 - r\Delta\xi, \eta - \eta_0 - s\Delta\eta\right) \right] \times$$

$$\exp\left( -i2\pi \frac{x\xi + y\eta}{\lambda Z} \right) d\xi d\eta . \qquad (6.5.3)$$

Replace in Eq. 6.5.3 the hologram window function $w(\xi,\eta)$ by its expression

$$w(\xi,\eta) = \int_{-\infty}^{\infty} \int_{-\infty}^{\infty} W(\overline{\xi},\overline{\eta}) \exp\left( i2\pi \frac{\xi\overline{\xi} + \eta\overline{\eta}}{\lambda Z} \right) d\overline{\xi}\, d\overline{\eta} \qquad (6.5.4)$$

through its Fourier spectrum $W(\overline{\xi},\overline{\eta})$ and obtain:

$$A_{rcstr}(x,y) = \int_{-\infty}^{\infty}\int_{-\infty}^{\infty}\int_{-\infty}^{\infty}\int_{-\infty}^{\infty} W(\overline{\xi},\overline{\eta}) \exp\left( i2\pi \frac{\xi\overline{\xi} + \eta\overline{\eta}}{\lambda Z} \right) d\overline{\xi}\, d\overline{\eta} \times$$

$$\sum_r \sum_s \Gamma_{r,s} h_{rec}\left(\xi - \xi_0 - r\Delta\xi, \eta - \eta_0 - s\Delta\eta\right) \exp\left( -i2\pi \frac{x\xi + y\eta}{\lambda Z} \right) d\xi d\eta =$$

$$\int_{-\infty}^{\infty}\int_{-\infty}^{\infty} W(\overline{\xi},\overline{\eta}) d\overline{\xi}\, d\overline{\eta} \sum_{r=-\infty}^{\infty} \sum_{s=-\infty}^{\infty} \Gamma_{r,s} \times$$

$$\int_{-\infty}^{\infty}\int_{-\infty}^{\infty} h_{rec}\left(\xi - \xi_0 - r\Delta\xi, \eta - \eta_0 - s\Delta\eta\right) \exp\left[ -i2\pi \frac{\left(x - \overline{\xi}\right)\xi + \left(y - \overline{\eta}\right)\eta}{\lambda Z} \right] d\xi d\eta =$$

$$\int_{-\infty}^{\infty}\int_{-\infty}^{\infty} W(\overline{\xi},\overline{\eta}) d\overline{\xi}\, d\overline{\eta} \sum_{r=-\infty}^{\infty} \sum_{s=-\infty}^{\infty} \Gamma_{r,s} \exp\left[ -i2\pi \frac{\left(x - \overline{\xi}\right)\left(\xi_0 + r\Delta\xi\right) + \left(y - \overline{\eta}\right)\left(\eta_0 + s\Delta\eta\right)}{\lambda Z} \right] \times$$

$$\int_{-\infty}^{\infty}\int_{-\infty}^{\infty} h_{rec}(\xi,\eta) \exp\left[ -i2\pi \frac{\left(x - \overline{\xi}\right)\xi + \left(y - \overline{\eta}\right)\eta}{\lambda Z} \right] d\xi d\eta . \qquad (6.5.5)$$

Replacing the last multiplicand in this equation by the frequency response, in coordinates $\left(x-\overline{\xi}, y-\overline{\eta}\right)$, of the hologram recording device:

$$H_{rec}\left(x-\overline{\xi}, y-\overline{\eta}\right) = \int\limits_{-\infty}^{\infty}\int\limits_{-\infty}^{\infty} h_{rec}\left(\xi,\eta\right)\exp\left[-i2\pi\frac{\left(x-\overline{\xi}\right)\xi+\left(y-\overline{\eta}\right)\eta}{\lambda Z}\right]d\xi\,d\eta, \quad (6.5.6)$$

obtain:

$$A_{rcstr}\left(x,y\right) = \int\limits_{-\infty}^{\infty}\int\limits_{-\infty}^{\infty} W\left(\overline{\xi},\overline{\eta}\right)H_{rec}\left(x-\overline{\xi}, y-\overline{\eta}\right)d\overline{\xi}\,d\overline{\eta} \times$$

$$\sum\limits_{r=-\infty}^{\infty}\sum\limits_{s=-\infty}^{\infty}\Gamma_{r,s}\exp\left[-i2\pi\frac{\left(x-\overline{\xi}\right)\left(\xi_0+r\Delta\xi\right)+\left(y-\overline{\eta}\right)\left(\eta_0+s\Delta\eta\right)}{\lambda Z}\right] =$$

$$\int\limits_{-\infty}^{\infty}\int\limits_{-\infty}^{\infty} W\left(\overline{\xi},\overline{\eta}\right)H_{rec}\left(x-\overline{\xi}, y-\overline{\eta}\right)\exp\left[-i2\pi\frac{\left(x-\overline{\xi}\right)\xi_0+\left(y-\overline{\eta}\right)\eta_0}{\lambda Z}\right] \times$$

$$\sum\limits_{r=-\infty}^{\infty}\sum\limits_{s=-\infty}^{\infty}\Gamma_{r,s}\exp\left[-i2\pi\frac{\left(x-\overline{\xi}\right)r\Delta\xi+\left(y-\overline{\eta}\right)s\Delta\eta}{\lambda Z}\right]d\overline{\xi}\,d\overline{\eta} =$$

$$\int\limits_{-\infty}^{\infty}\int\limits_{-\infty}^{\infty} W\left(x-\overline{\xi}, y-\overline{\eta}\right)H_{rec}\left(\overline{\xi},\overline{\eta}\right)\exp\left(-i2\pi\frac{\overline{\xi}\xi_0+\overline{\eta}\eta_0}{\lambda Z}\right) \times$$

$$\sum\limits_{r=-\infty}^{\infty}\sum\limits_{s=-\infty}^{\infty}\Gamma_{r,s}\exp\left(-i2\pi\frac{\overline{\xi}\Delta\xi r+\overline{\eta}\Delta\eta s}{\lambda Z}\right)d\overline{\xi}\,d\overline{\eta}. \quad (6.5.7)$$

Denoting

$$\tilde{H}_{rec}^{(\xi_0,\eta_0)}\left(\overline{\xi},\overline{\eta}\right) = H_{rec}\left(\overline{\xi},\overline{\eta}\right)\exp\left(-i2\pi\frac{\overline{\xi}\xi_0+\overline{\eta}\eta_0}{\lambda Z}\right) \quad (6.5.8)$$

and replacing $\Gamma_{r,s}$ by its expression 6.5.1 through object wave front samples $\overline{A}_{k,l}^{(kf)}$, results in:

$$A_{rcstr}\left(x,y\right) = \int\limits_{-\infty}^{\infty}\int\limits_{-\infty}^{\infty} W\left(x-\overline{\xi}, y-\overline{\eta}\right)\tilde{H}_{rec}\left(\overline{\xi},\overline{\eta}\right) \times$$

$$\sum\limits_{r=-\infty}^{\infty}\sum\limits_{s=-\infty}^{\infty}\sum\limits_{k=0}^{N_1-1}\sum\limits_{l=0}^{N_2-1}\overline{A}_{k,l}^{(kf)}\exp\left\{i2\pi\left[\frac{\left(k+u\right)\left(r+p\right)}{N_1}+\frac{\left(l+v\right)\left(s+q\right)}{N_2}\right]\right\} \times$$

$$\exp\left(-i2\pi\frac{\overline{\xi}\Delta\xi r+\overline{\eta}\Delta\eta s}{\lambda Z}\right)d\overline{\xi}\,d\overline{\eta}=$$

$$\int_{-\infty}^{\infty}\int_{-\infty}^{\infty}W\left(x-\overline{\xi},y-\overline{\eta}\right)\tilde{H}_{rec}\left(\overline{\xi},\overline{\eta}\right)\sum_{k=0}^{N_1-1}\sum_{l=0}^{N_2-1}\overline{A}_{k,l}^{(kf)}\exp\left[i2\pi\left(\frac{k+u}{N_1}p+\frac{l+v}{N_2}q\right)\right]\times$$

$$\sum_{r=-\infty}^{\infty}\sum_{s=-\infty}^{\infty}\exp\left[i2\pi\left(\frac{k+u}{N_1}r+\frac{l+v}{N_2}s\right)\right]\exp\left(-i2\pi\frac{\overline{\xi}\Delta\xi r+\overline{\eta}\Delta\eta s}{\lambda Z}\right)d\overline{\xi}\,d\overline{\eta}=$$

$$\int_{-\infty}^{\infty}\int_{-\infty}^{\infty}W\left(x-\overline{\xi},y-\overline{\eta}\right)\tilde{H}_{rec}\left(\overline{\xi},\overline{\eta}\right)\sum_{k=0}^{N_1-1}\sum_{l=0}^{N_2-1}\overline{A}_{k,l}^{(kf)}\exp\left[i2\pi\left(\frac{k+u}{N_1}p+\frac{l+v}{N_2}q\right)\right]\times$$

$$\sum_{r=-\infty}^{\infty}\exp\left[-i2\pi\left(\frac{\overline{\xi}\Delta\xi}{\lambda Z}-\frac{k+u}{N_2}\right)r\right]\sum_{s=-\infty}^{\infty}\exp\left[-i2\pi\left(\frac{\overline{\eta}\Delta\eta}{\lambda Z}-\frac{l+v}{N_2}\right)s\right]d\overline{\xi}\,d\overline{\eta}. \qquad (6.5.9)$$

The last two multiplicands in Eq. 6.5.9 can be modified using the Poisson summation formula (Eq. 6.2.10). Then obtain:

$$A_{rcstr}\left(x,y\right)=\int_{-\infty}^{\infty}\int_{-\infty}^{\infty}W\left(x-\overline{\xi},y-\overline{\eta}\right)\tilde{H}_{rec}\left(\overline{\xi},\overline{\eta}\right)\sum_{k=0}^{N-1}\sum_{l=0}^{N_2-1}\overline{A}_{k,l}^{(kf)}\exp\left[i2\pi\left(\frac{k+u}{2N_1}p+\frac{l+v}{N_2}q\right)\right]\times$$

$$\sum_{do_x=-\infty}^{\infty}\sum_{do_y=-\infty}^{\infty}\delta\left(\frac{\overline{\xi}\Delta\xi}{\lambda Z}-\frac{k+u}{N_2}-do_x\right)\delta\left(\frac{\overline{\eta}\Delta\eta}{\lambda Z}-\frac{l+v}{N_2}-do_y\right)d\overline{\xi}\,d\overline{\eta}=$$

$$\int_{-\infty}^{\infty}\int_{-\infty}^{\infty}W\left(x-\overline{\xi},y-\overline{\eta}\right)\tilde{H}_{rec}\left(\overline{\xi},\overline{\eta}\right)\sum_{k=0}^{N_1-1}\sum_{l=0}^{N_2-1}\overline{A}_{k,l}^{(kf)}\exp\left[i2\pi\left(\frac{k+u}{2N_1}p+\frac{l+v}{N_2}q\right)\right]\times$$

$$\sum_{r=-\infty}^{\infty}\sum_{s=-\infty}^{\infty}\delta\left(\frac{\overline{\xi}\Delta\xi}{\lambda Z}-\frac{k+u}{N_2}-do_x\right)\delta\left(\frac{\overline{\eta}\Delta\eta}{\lambda Z}-\frac{l+v}{N_2}-do_y\right)d\overline{\xi}\,d\overline{\eta}=$$

$$\sum_{k=0}^{N_1-1}\sum_{l=0}^{N_2-1}\overline{A}_{k,l}^{(kf)}\exp\left[i2\pi\left(\frac{k+u}{2N_1}p+\frac{l+v}{N_2}q\right)\right]\times$$

$$\sum_{do_x=-\infty}^{\infty}\sum_{do_y=-\infty}^{\infty}\tilde{H}_{rec}\left[\left(\frac{k+u}{N_1}+do_x\right)\frac{\lambda Z}{\Delta\xi},\left(\frac{l+v}{N_2}+do_y\right)\frac{\lambda Z}{\Delta\eta}\right]\times$$

$$W\left(x-\frac{k+u}{N_1}\frac{\lambda Z}{\Delta\xi}-do_x\frac{\lambda Z}{\Delta\xi};y-\frac{l+v}{N_2}\frac{\lambda Z}{\Delta\eta}-do_y\frac{\lambda Z}{\Delta\eta}\right). \qquad (6.5.10)$$

With selection $p=0$ and $q=0$, obtain finally

$$A_{rcstr}\left(x,y\right)=\sum_{k=0}^{N_1-1}\sum_{l=0}^{N_2-1}\sum_{do_x=-\infty}^{\infty}\sum_{do_y=-\infty}^{\infty}\overline{A}_{k,l}^{(kf)}\tilde{H}_{rec}\left[\left(\frac{k+u}{N_1}+do_x\right)\frac{\lambda Z}{\Delta\xi},\left(\frac{l+v}{N_2}+do_y\right)\frac{\lambda Z}{\Delta\eta}\right]\times$$

$$W\left[x-\left(\frac{k+u}{N_1}+do_x\right)\frac{\lambda Z}{\Delta\xi};y-\left(\frac{l+v}{N_2}+do_y\right)\frac{\lambda Z}{\Delta\eta}\right]=\sum_{do_x=-\infty}^{\infty}\sum_{do_y=-\infty}^{\infty}\overline{A}_{k,l}^{(kf)}\left(x,do_x;y,do_y\right)$$

$$(6.5.11)$$

where it is denoted:

$$\overline{A}_{k,l}^{(kf)}\left(x,do_x;y,do_y\right)=\sum_{k=0}^{N_1-1}\sum_{l=0}^{N_2-1}\tilde{H}_{rec}\left[\left(\frac{k+u}{N_1}-do_x\right)\frac{\lambda Z}{\Delta\xi},\left(\frac{l+v}{N_2}-do_y\right)\frac{\lambda Z}{\Delta\eta}\right]\overline{A}_{k,l}^{(kf)}\times$$

$$W\left(x-\frac{k+u}{N_1}\frac{\lambda Z}{\Delta\xi}-do_x\frac{\lambda Z}{\Delta\xi};y-\frac{l+v}{N_2}\frac{\lambda Z}{\Delta\eta}-do_y\frac{\lambda Z}{\Delta\eta}\right).\qquad(6.5.12)$$

Eq. 6.5.12 has a clear physical interpretation similar to that for the case of holograms generated using the symmetrization method.

- Object wave front is reconstructed in a number of diffraction orders $\left\{do_x,do_y\right\}$

- In the diffraction order $\left(do_x,do_y\right)$, the reconstructed wave front is a result of interpolation of samples $\left\{\overline{A}_{k,l}^{(kf)}\right\}$ of the object wave front modified for kinoform encoding, with an interpolation kernel $W\left(x;y\right)$, which is Fourier transform of the recorded hologram window function $w\left(\xi,\eta\right)$, those samples being weighted by samples of the frequency response $\tilde{H}_{rec}\left(x,y\right)$ of the hologram recording device taken at coordinates $\left\{x=\left(\frac{k+u}{N_1}+do_x\right)\frac{\lambda Z}{\Delta\xi},y=\left(\frac{l+v}{N_2}+do_y\right)\frac{\lambda Z}{\Delta\eta}\right\}$.

This interpretation is illustrated in Fig. 6.6.

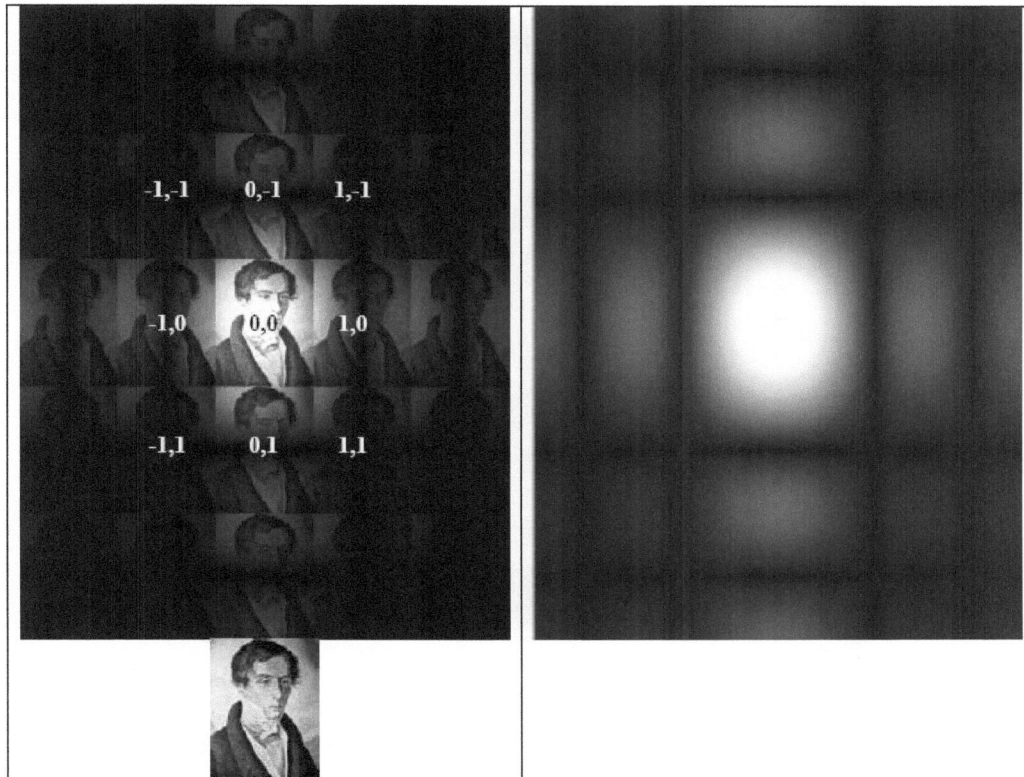

**Fig. 6.6.** Left side: computer simulation of optical reconstruction of computer-generated kinoform of the image shown at the bottom; numbers indicate diffraction order indices. Right side: reconstruction masking function for a rectangular hologram recording aperture of $\Delta\xi \times \Delta\eta$ size.

## Appendix. Derivation of Eq. 6.2.10.

Let

$$\tilde{\Gamma}\left(\xi,\eta\right) = w\left(\xi,\eta\right)\sum_{r}\sum_{s}\left(\Gamma_{r,s}+b\right)h_{rec}\left(\xi-\xi_0-r\Delta\xi,\eta-\eta_0-s\Delta\eta\right); \qquad (A6.1)$$

be a recorded hologram. Then the reconstructed waveform can be found as integral Fourier transform:

$$A_{rcstr}\left(x,y\right) = \int\limits_{-\infty}^{\infty}\int\limits_{-\infty}^{\infty}\left[w\left(\xi,\eta\right)\sum_{r}\sum_{s}\left(\Gamma_{r,s}+b\right)h_{rec}\left(\xi-\xi_0-r\Delta\xi,\eta-\eta_0-s\Delta\eta\right)\right]\times$$

$$\exp\left(-i2\pi\frac{x\xi+y\eta}{\lambda Z}\right)d\xi d\eta = \int\limits_{-\infty}^{\infty}\int\limits_{-\infty}^{\infty}\int\limits_{-\infty}^{\infty}\int\limits_{-\infty}^{\infty} W\left(\overline{\xi},\overline{\eta}\right)\exp\left(i2\pi\frac{\xi\overline{\xi}+\eta\overline{\eta}}{\lambda Z}\right)d\overline{\xi}\,d\overline{\eta}\times$$

$$\sum_{r}\sum_{s}\left(\Gamma_{r,s}+b\right)h_{rec}\left(\xi-\xi_0-r\Delta\xi,\eta-\eta_0-s\Delta\eta\right)\exp\left(-i2\pi\frac{x\xi+y\eta}{\lambda Z}\right)d\xi d\eta =$$

$$\int\limits_{-\infty}^{\infty}\int\limits_{-\infty}^{\infty} W\left(\overline{\xi},\overline{\eta}\right)d\overline{\xi}\,d\overline{\eta}\sum_{r=-\infty}^{\infty}\sum_{s=-\infty}^{\infty}\left(\Gamma_{r,s}+b\right)\times$$

$$\int\limits_{-\infty}^{\infty}\int\limits_{-\infty}^{\infty} h_{rec}\left(\xi-\xi_0-r\Delta\xi,\eta-\eta_0-s\Delta\eta\right)\exp\left[-i2\pi\frac{\left(x-\overline{\xi}\right)\xi+\left(y-\overline{\eta}\right)\eta}{\lambda Z}\right]d\xi\,d\eta=$$

$$\int\limits_{-\infty}^{\infty}\int\limits_{-\infty}^{\infty} W\left(\overline{\xi},\overline{\eta}\right)d\overline{\xi}\,d\overline{\eta}\sum_{r=-\infty}^{\infty}\sum_{s=-\infty}^{\infty}\left(\Gamma_{r,s}+b\right)\exp\left[-i2\pi\frac{\left(x-\overline{\xi}\right)\left(\xi_0+r\Delta\xi\right)+\left(y-\overline{\eta}\right)\left(\eta_0+s\Delta\eta\right)}{\lambda Z}\right]\times$$

$$\int\limits_{-\infty}^{\infty}\int\limits_{-\infty}^{\infty} h_{rec}\left(\xi,\eta\right)\exp\left[-i2\pi\frac{\left(x-\overline{\xi}\right)\xi+\left(y-\overline{\eta}\right)\eta}{\lambda Z}\right]d\xi\,d\eta=$$

$$\int\limits_{-\infty}^{\infty}\int\limits_{-\infty}^{\infty} W\left(\overline{\xi},\overline{\eta}\right)d\overline{\xi}\,d\overline{\eta}\sum_{r=-\infty}^{\infty}\sum_{s=-\infty}^{\infty}\left(\Gamma_{r,s}+b\right)\exp\left[-i2\pi\frac{\left(x-\overline{\xi}\right)\left(\xi_0+r\Delta\xi\right)+\left(y-\overline{\eta}\right)\left(\eta_0+s\Delta\eta\right)}{\lambda Z}\right]\times$$

$$H_{rec}\left(x-\overline{\xi},y-\overline{\eta}\right)\ ,\tag{A6.2}$$

where

$$H_{rec}\left(x-\overline{\xi},y-\overline{\eta}\right)=\int\limits_{-\infty}^{\infty}\int\limits_{-\infty}^{\infty} h_{rec}\left(\xi,\eta\right)\exp\left[-i2\pi\frac{\left(x-\overline{\xi}\right)\xi+\left(y-\overline{\eta}\right)\eta}{\lambda Z}\right]d\xi\,d\eta\ .\tag{A6.3}$$

With this replacement:

$$A_{rcstr}\left(x,y\right)=\int\limits_{-\infty}^{\infty}\int\limits_{-\infty}^{\infty} W\left(\overline{\xi},\overline{\eta}\right)H_{rec}\left(x-\overline{\xi},y-\overline{\eta}\right)\exp\left[-i2\pi\frac{\left(x-\overline{\xi}\right)\xi_0+\left(y-\overline{\eta}\right)\eta_0}{\lambda Z}\right]\times$$

$$\sum_{r=-\infty}^{\infty}\sum_{s=-\infty}^{\infty}\left(\Gamma_{r,s}+b\right)\exp\left[-i2\pi\frac{\left(x-\overline{\xi}\right)r\Delta\xi+\left(y-\overline{\eta}\right)s\Delta\eta}{\lambda Z}\right]d\overline{\xi}\,d\overline{\eta}$$

$$\int\limits_{-\infty}^{\infty}\int\limits_{-\infty}^{\infty} W\left(x-\overline{\xi},y-\overline{\eta}\right)H_{rec}\left(\overline{\xi},\overline{\eta}\right)\exp\left(-i2\pi\frac{\overline{\xi}\xi_0+\overline{\eta}\eta_0}{\lambda Z}\right)\times$$

$$\sum_{r=-\infty}^{\infty}\sum_{s=-\infty}^{\infty}\left(\Gamma_{r,s}+b\right)\exp\left(-i2\pi\frac{\overline{\xi}\Delta\xi r+\overline{\eta}\Delta\eta s}{\lambda Z}\right)d\overline{\xi}\,d\overline{\eta}\ .\tag{A6.4}$$

Consider the first term

$$A_{rcstr}^{(\Gamma)}\left(x,y\right)=\int\limits_{-\infty}^{\infty}\int\limits_{-\infty}^{\infty} W\left(x-\overline{\xi},y-\overline{\eta}\right)H_{rec}\left(\overline{\xi},\overline{\eta}\right)\exp\left(-i2\pi\frac{\overline{\xi}\xi_0+\overline{\eta}\eta_0}{\lambda Z}\right)\times$$

$$\sum_{r=-\infty}^{\infty}\sum_{s=-\infty}^{\infty}\Gamma_{r,s}\exp\left(-i2\pi\frac{\overline{\xi}\Delta\xi r+\overline{\eta}\Delta\eta s}{\lambda Z}\right)d\overline{\xi}\,d\overline{\eta}\ ;\tag{A6.5}$$

Denoting

$$\tilde{H}_{rec}^{(\xi_0,\eta_0)}\left(\overline{\xi},\overline{\eta}\right)=H_{rec}\left(\overline{\xi},\overline{\eta}\right)\exp\left(-i2\pi\frac{\overline{\xi}\xi_0+\overline{\eta}\eta_0}{\lambda Z}\right) \tag{A6.6}$$

and substituting it into Eq. A6.5

$$\Gamma_{r,s}=\sum_{k=0}^{2N_1-1}\sum_{l=0}^{N_2-1}\overline{A}_{k,l}^{(o)}\exp\left\{i2\pi\left[\frac{(k+u)(r+p)}{2N_1}+\frac{(l+v)(s+q)}{N_2}\right]\right\}, \tag{A6.7}$$

obtain:

$$A_{rcstr}^{(\Gamma)}\left(x,y\right)=\int_{-\infty}^{\infty}\int_{-\infty}^{\infty}W\left(x-\overline{\xi},y-\overline{\eta}\right)\tilde{H}_{rec}\left(\overline{\xi},\overline{\eta}\right)\times$$

$$\sum_{r=-\infty}^{\infty}\sum_{s=-\infty}^{\infty}\sum_{k=0}^{2N_1-1}\sum_{l=0}^{N_2-1}\overline{A}_{k,l}^{(o)}\exp\left\{i2\pi\left[\frac{(k+u)(r+p)}{2N_1}+\frac{(l+v)(s+q)}{N_2}\right]\right\}\times$$

$$\exp\left(-i2\pi\frac{\overline{\xi}\Delta\xi r+\overline{\eta}\Delta\eta s}{\lambda Z}\right)d\overline{\xi}\,d\overline{\eta}=$$

$$\int_{-\infty}^{\infty}\int_{-\infty}^{\infty}W\left(x-\overline{\xi},y-\overline{\eta}\right)\tilde{H}_{rec}\left(\overline{\xi},\overline{\eta}\right)\sum_{k=0}^{2N_1-1}\sum_{l=0}^{N_2-1}\overline{A}_{k,l}^{(o)}\exp\left[i2\pi\left(\frac{k+u}{2N_1}p+\frac{l+v}{N_2}q\right)\right]\times$$

$$\sum_{r=-\infty}^{\infty}\sum_{s=-\infty}^{\infty}\exp\left[i2\pi\left(\frac{k+u}{2N_1}r+\frac{l+v}{N_2}s\right)\right]\exp\left(-i2\pi\frac{\overline{\xi}\Delta\xi r+\overline{\eta}\Delta\eta s}{\lambda Z}\right)d\overline{\xi}\,d\overline{\eta}=$$

$$\int_{-\infty}^{\infty}\int_{-\infty}^{\infty}W\left(x-\overline{\xi},y-\overline{\eta}\right)\tilde{H}_{rec}\left(\overline{\xi},\overline{\eta}\right)\sum_{k=0}^{2N_1-1}\sum_{l=0}^{N_2-1}\overline{A}_{k,l}^{(o)}\exp\left[i2\pi\left(\frac{k+u}{2N_1}p+\frac{l+v}{N_2}q\right)\right]\times$$

$$\sum_{r=-\infty}^{\infty}\exp\left[-i2\pi\left(\frac{\overline{\xi}\Delta\xi}{\lambda Z}-\frac{k+u}{N_2}\right)r\right]\sum_{s=-\infty}^{\infty}\exp\left[-i2\pi\left(\frac{\overline{\eta}\Delta\eta}{\lambda Z}-\frac{l+v}{N_2}\right)s\right]d\overline{\xi}\,d\overline{\eta}. \tag{A6.8}$$

Now use Poisson summation formula:

$$\sum_{r=-\infty}^{\infty}\exp\left(-i2\pi fr\right)=\sum_{do=-\infty}^{\infty}\delta\left(f-do\right), \tag{A6.9}$$

for the last line of the Eq. A6.8 and obtain

$$A_{rcstr}^{(\Gamma)}\left(x,y\right)=$$

$$\int_{-\infty}^{\infty}\int_{-\infty}^{\infty}W\left(x-\overline{\xi},y-\overline{\eta}\right)\tilde{H}_{rec}\left(\overline{\xi},\overline{\eta}\right)\sum_{k=0}^{2N_1-1}\sum_{l=0}^{N_2-1}\overline{A}_{k,l}^{(o)}\exp\left[i2\pi\left(\frac{k+u}{2N_1}p+\frac{l+v}{N_2}q\right)\right]\times$$

$$\sum_{r=-\infty}^{\infty}\sum_{s=-\infty}^{\infty}\delta\left(\frac{\overline{\xi}\Delta\xi}{\lambda Z}-\frac{k+u}{N_2}-do_x\right)\delta\left(\frac{\overline{\eta}\Delta\eta}{\lambda Z}-\frac{l+v}{N_2}-do_y\right)d\overline{\xi}\,d\overline{\eta}=$$

$$\int\limits_{-\infty}^{\infty}\int\limits_{-\infty}^{\infty} W\left(x-\bar{\xi},y-\bar{\eta}\right)\tilde{H}_{rec}\left(\bar{\xi},\bar{\eta}\right)\sum_{k=0}^{2N_1-1}\sum_{l=0}^{N_2-1}\overline{A}_{k,l}^{(o)}\exp\left[i2\pi\left(\frac{k+u}{2N_1}p+\frac{l+v}{N_2}q\right)\right]\times$$

$$\sum_{r=-\infty}^{\infty}\sum_{s=-\infty}^{\infty}\delta\left(\frac{\bar{\xi}\Delta\xi}{\lambda Z}-\frac{k+u}{N_2}-do_x\right)\delta\left(\frac{\bar{\eta}\Delta\eta}{\lambda Z}-\frac{l+v}{N_2}-do_y\right)d\bar{\xi}\,d\bar{\eta}=$$

$$\sum_{k=0}^{2N_1-1}\sum_{l=0}^{N_2-1}\overline{A}_{k,l}^{(o)}\exp\left[i2\pi\left(\frac{k+u}{2N_1}p+\frac{l+v}{N_2}q\right)\right]\times$$

$$\sum_{do_x=-\infty}^{\infty}\sum_{do_y=-\infty}^{\infty}\tilde{H}_{rec}\left[\left(\frac{k+u}{N_1}-do_x\right)\frac{\lambda Z}{\Delta\xi},\left(\frac{l+v}{N_2}-do_y\right)\frac{\lambda Z}{\Delta\xi}\right]\times$$

$$W\left(x-\frac{k+u}{N_1}\frac{\lambda Z}{\Delta\xi}-do_x\frac{\lambda Z}{\Delta\xi};x-\frac{l+v}{N_2}\frac{\lambda Z}{\Delta\xi}-do_y\frac{\lambda Z}{\Delta\xi}\right). \qquad (A6.10)$$

Set shift parameters $p$ and $q$ to $p=0$ and $q=0$. With this setting obtain:

$$A_{rcstr}^{(\Gamma)}\left(x,y\right)\propto\sum_{k=0}^{2N_1-1}\sum_{l=0}^{N_2-1}\sum_{do_x=-\infty}^{\infty}\sum_{do_y=-\infty}^{\infty}\tilde{H}_{rec}\left[\left(\frac{k+u}{N_1}-do_x\right)\frac{\lambda Z}{\Delta\xi},\left(\frac{l+v}{N_2}-do_y\right)\frac{\lambda Z}{\Delta\xi}\right]\overline{A}_{k,l}^{(o)}\times$$

$$W\left(x-\frac{k+u}{N_1}\frac{\lambda Z}{\Delta\xi}-do_x\frac{\lambda Z}{\Delta\xi};x-\frac{l+v}{N_2}\frac{\lambda Z}{\Delta\xi}-do_y\frac{\lambda Z}{\Delta\xi}\right). \qquad (A6.11)$$

Now consider the second term and obtain similarly:

$$A_{rcstr}^{(b)}\left(x,y\right)=b\int\limits_{-\infty}^{\infty}\int\limits_{-\infty}^{\infty} W\left(x-\bar{\xi},y-\bar{\eta}\right)H_{rec}\left(\bar{\xi},\bar{\eta}\right)\exp\left(-i2\pi\frac{\bar{\xi}\xi_0+\bar{\eta}\eta_0}{\lambda Z}\right)\times$$

$$\sum_{r=-\infty}^{\infty}\sum_{s=-\infty}^{\infty}\exp\left(-i2\pi\frac{\bar{\xi}\Delta\xi r+\bar{\eta}\Delta\eta s}{\lambda Z}\right)d\bar{\xi}\,d\bar{\eta}=$$

$$b\int\limits_{-\infty}^{\infty}\int\limits_{-\infty}^{\infty} W\left(x-\bar{\xi},y-\bar{\eta}\right)H_{rec}\left(\bar{\xi},\bar{\eta}\right)\exp\left(-i2\pi\frac{\bar{\xi}\xi_0+\bar{\eta}\eta_0}{\lambda Z}\right)\times$$

$$\sum_{r=-\infty}^{\infty}\exp\left(-i2\pi\frac{\bar{\xi}\Delta\xi}{\lambda Z}r\right)\sum_{s=-\infty}^{\infty}\exp\left(-i2\pi\frac{\bar{\eta}\Delta\eta}{\lambda Z}s\right)d\bar{\xi}\,d\bar{\eta}=$$

$$b\int\limits_{-\infty}^{\infty}\int\limits_{-\infty}^{\infty} W\left(x-\bar{\xi},y-\bar{\eta}\right)\tilde{H}_{rec}^{(\xi_0,\eta_0)}\left(\bar{\xi},\bar{\eta}\right)\sum_{do_x=-\infty}^{\infty}\delta\left(\frac{\bar{\xi}\Delta\xi}{\lambda Z}-do_x\right)\sum_{do_y=-\infty}^{\infty}\delta\left(\frac{\bar{\eta}\Delta\eta}{\lambda Z}-do_y\right)d\bar{\xi}\,d\bar{\eta}=$$

$$b\sum_{do_x=-\infty}^{\infty}\sum_{do_y=-\infty}^{\infty}\int\limits_{-\infty}^{\infty}\int\limits_{-\infty}^{\infty} W\left(x-\bar{\xi},y-\bar{\eta}\right)\tilde{H}_{rec}^{(\xi_0,\eta_0)}\left(\bar{\xi},\bar{\eta}\right)\delta\left(\frac{\bar{\xi}\Delta\xi}{\lambda Z}-do_x\right)\delta\left(\frac{\bar{\eta}\Delta\eta}{\lambda Z}-do_y\right)d\bar{\xi}\,d\bar{\eta}=$$

$$b\sum_{do_x=-\infty}^{\infty}\sum_{do_y=-\infty}^{\infty} W\left(x-\frac{\lambda Z}{\Delta\xi}do_x,y-\frac{\lambda Z}{\Delta\xi}do_y\right)\tilde{H}_{rec}^{(\xi_0,\eta_0)}\left(\frac{\lambda Z}{\Delta\xi}do_x,\frac{\lambda Z}{\Delta\xi}do_y\right). \qquad (A6.12)$$

Thus obtain finally for the reconstructed object wave front:

$$A_{rcstr}\left(x,y\right) = A_{rcstr}^{(\Gamma)}\left(x,y\right) + A_{rcstr}^{(b)}\left(x,y\right) \propto$$

$$\sum_{do_x=-\infty}^{\infty}\sum_{do_y=-\infty}^{\infty}\sum_{k=0}^{2N_1-1}\sum_{l=0}^{N_2-1}\tilde{H}_{rec}\left[\left(\frac{k+u}{N_1}-do_x\right)\frac{\lambda Z}{\Delta\xi},\left(\frac{l+v}{N_2}-do_y\right)\frac{\lambda Z}{\Delta\xi}\right]\overline{A}_{k,l}^{(o)}\times$$

$$W\left(x-\frac{k+u}{N_1}\frac{\lambda Z}{\Delta\xi}-do_x\frac{\lambda Z}{\Delta\xi};x-\frac{l+v}{N_2}\frac{\lambda Z}{\Delta\xi}-do_y\frac{\lambda Z}{\Delta\xi}\right)+$$

$$b\sum_{do_x=-\infty}^{\infty}\sum_{do_y=-\infty}^{\infty}W\left(x-\frac{\lambda Z}{\Delta\xi}do_x,y-\frac{\lambda Z}{\Delta\xi}do_y\right)\tilde{H}_{rec}^{(\xi_0,\eta_0)}\left(\frac{\lambda Z}{\Delta\xi}do_x,\frac{\lambda Z}{\Delta\xi}do_y\right)=$$

$$\sum_{do_x=-\infty}^{\infty}\sum_{do_y=-\infty}^{\infty}\tilde{\overline{A}}^{(o)}\left(x,do_x;y,do_y\right)+$$

$$b\sum_{do_x=-\infty}^{\infty}\sum_{do_y=-\infty}^{\infty}W\left(x-\frac{\lambda Z}{\Delta\xi}do_x,y-\frac{\lambda Z}{\Delta\xi}do_y\right)\tilde{H}_{rec}^{(\xi_0,\eta_0)}\left(\frac{\lambda Z}{\Delta\xi}do_x,\frac{\lambda Z}{\Delta\xi}do_y\right), \qquad (A6.13)$$

where it is denoted:

$$\tilde{\overline{A}}^{(o)}\left(x,do_x;y,do_y\right) = \sum_{k=0}^{2N_1-1}\sum_{l=0}^{N_2-1}\tilde{H}_{rec}\left[\left(\frac{k+u}{N_1}-do_x\right)\frac{\lambda Z}{\Delta\xi},\left(\frac{l+v}{N_2}-do_y\right)\frac{\lambda Z}{\Delta\xi}\right]\overline{A}_{k,l}^{(o)}\times$$

$$W\left(x-\frac{k+u}{N_1}\frac{\lambda Z}{\Delta\xi}-do_x\frac{\lambda Z}{\Delta\xi};x-\frac{l+v}{N_2}\frac{\lambda Z}{\Delta\xi}-do_y\frac{\lambda Z}{\Delta\xi}\right). \qquad (A6.14)$$

# 7 Computer-Generated Holograms and Optical Information Processing

**Abstract:** A remarkable property of lenses and parabolic mirrors is their ability of performing, in parallel and at the speed of light, Fourier transform of input wave fronts and to act as "chirp"- spatial light modulator. This basic property enables creating optical information processing systems for implementation of optical Fourier analysis, image convolution and correlation. In this chapter, basic properties of lenses and parabolic mirrors relevant to optical information processing are briefly explained, elements of the theory of optical correlators for reliable target location in images are an introduced and their different implementations of optical correlators are described.

## 7.1 Principles of optical information processing

### 7.1.1 Lens as a spatial light modulator

Consider schematic diagram presented in Figs. 7.1.

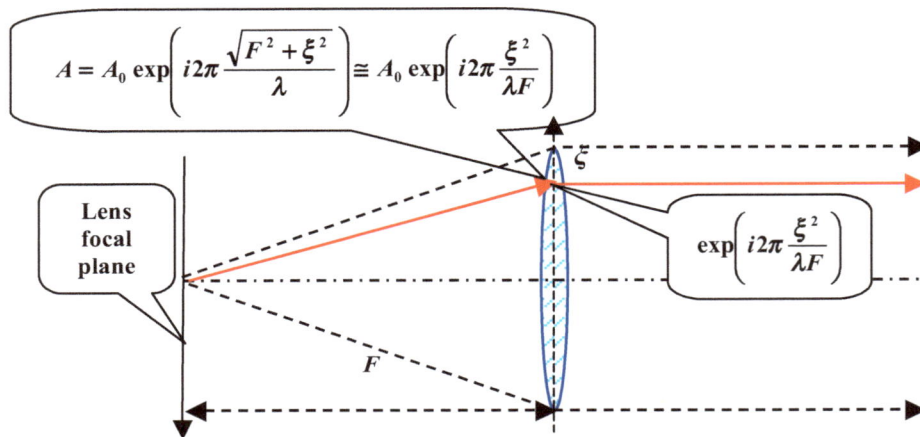

**Fig. 7.1.** Lens as a "chirp" spatial light modulator that converts spherical wave front to plane one

One can see from the figure that coherent light from a point source at the lens focal point arrives at a point with coordinate $\xi$ in the lens plane with a phase shift $\pi\sqrt{F^2+\xi^2}/\lambda$ with respect to the source phase, where $F$ is the lens focal distance and $\lambda$ is the light wave length. If the size of the lens is sufficiently small than the lens focal distance, this phase shift as a function of coordinate $\xi$ can be approximated as $\pi\xi^2/\lambda F$. The fundamental property of lenses is that lenses convert a spherical wave front propagating from a point source located at the lens focal plane into a plane wave front. This implies that the lens in the set up of Fig. 7.1 acts as a spatial light modulator with transfer function $\exp(-i\pi\xi^2/\lambda F)$. Exponential function of quadratic imaginary variable is called *"chirp"-function*, hence the name *"chirp modulator"* frequently used to characterize this property of lens as a spatial light modulator.

## 7.1.2   Lenses and parabolic mirrors as a Fourier transformers

Schematic diagram in Fig. 7.2 illustrates the property of lenses to perform integral Fourier transform. Consider a spherical wave front propagating from a point source at coordinate $x$ in the frontal focal plane of the lens. The lens converts this wave front into a plane wave front propagating to the rear focal plane of the lens with a slope $x/F$, so that at a point with coordinate $f$ at the lens rear focal plane it has a phase shift $2\pi x/\lambda F$ with respect to its phase on the optical axis at point $f = 0$. This implies that the point spread function $PSF(x, f)$ that describes wave propagation between frontal and rear focal planes of the lens is $\exp(i2\pi\, fx/\lambda F)$, which is the kernel of the integral Fourier transform.

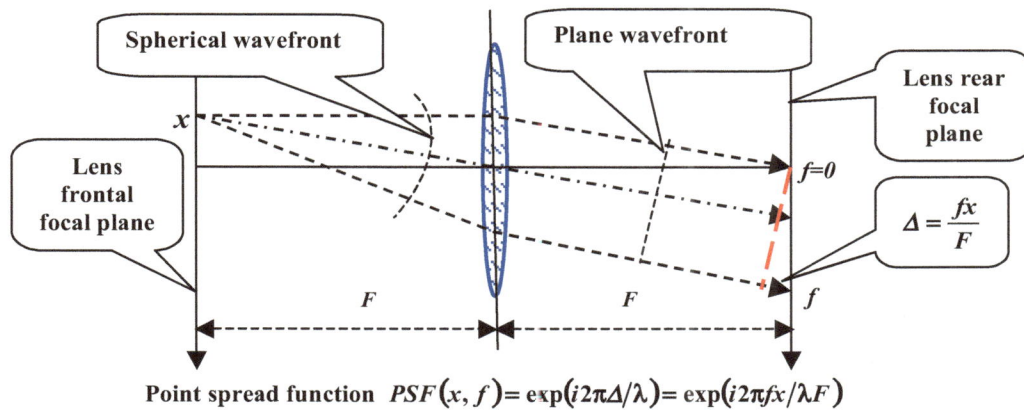

$$\text{Point spread function } PSF(x, f) = \exp(i2\pi\Delta/\lambda) = \exp(i2\pi fx/\lambda F)$$

**Fig. 7.2.** Lens as Fourier processor that converts amplitude distribution of the wave front in its fore focal plane into distribution of its Fourier transform in the rear focal plane

Parabolic mirrors illustrated in Fig. 7.3, similarly to lenses, also convert a spherical wavefront emanating from the mirror focal point into a plane wavefront and, reciprocally, a plane incoming wavefront into a spherical one converging in the focal point.

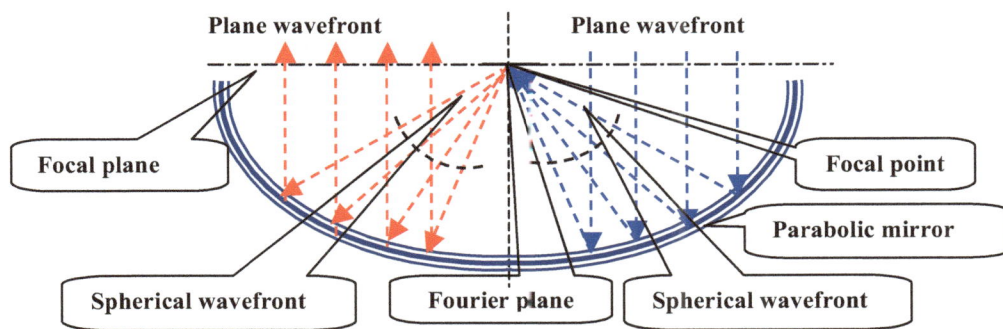

**Fig. 7.3.** Parabolic mirror as a "chirp" spatial light modulator and Fourier transformer

Therefore parabolic mirrors also act with respect to incoming wavefronts of coherent light as Fourier transformers and chirp spatial light modulators.

### 7.1.3 *Electro-optical image processing systems*

Schematic diagram of a classical electro-optical image processing system that make use of described properties of lenses as Fourier transformers is presented in Fig. 7.4. Fig. 7.5 shows a schematic diagram of a system in which lenses are replaced by a parabolic mirror as a Fourier transformer. This replacement allows making the system more compact.

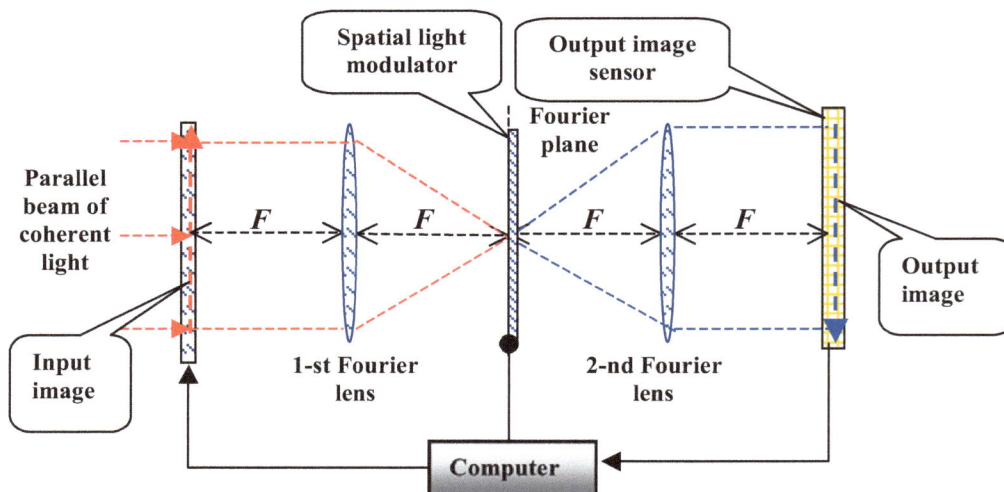

**Fig. 7.4.** F electro-optical Fourier processor for image spatial filtering

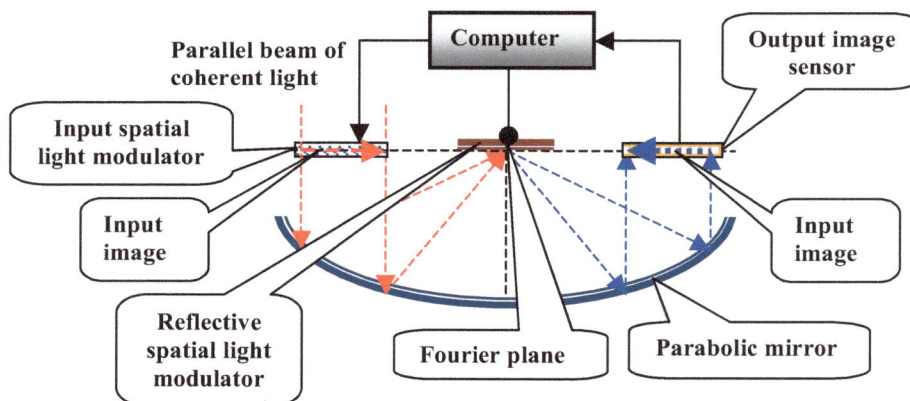

**Fig. 7.5.** Parabolic mirror electro-optical Fourier processor for image spatial filtering

The systems work as following. Input images recorded on the input spatial light modulator are illuminated by a parallel beam of laser light and Fourier transformed by the fiorst Fourier lens in the system of Fig. 7.4 or by the left side of the parabolic mirror in the system of Fig. 7.5. Image spectrum in the system Fourier plane is modulated by its amplitude and phase by a spatial filter recorded on a spatial light modulator installed in the Fourier plane. The modified by the spatial filter image spectrum is then Fourier transformed by the second Fourier lens in the lens system or, correspondingly, by the right part of the parabolic mirror to form an output image on the output image sensor. In the parabolic mirror system, spatial light modulator installed in the mirror focal plane has to be a reflective one. In this system, input image plane, output image plane and Fourier plane coincide, which makes the system much more compact than the lens based system.

In both systems, recording input image and spatial filter as well as reading output is controlled by a computer. As spatial filters, computer-generated holograms can be used.

Such image processing systems can, in principle, implement image convolution with any kernel specified by the spatial filter in the system Fourier plane. In particular, they can be used to compute correlation of input images with template images. This is an application that attracted most attention of researches ([1-4]). We will address it in the section that follows.

## 7.2   Optical correlators for target location: elements of the theory

Consider the task of target location in images. Let $b(x,y)$ be an observed image that contains a target image $a(x-x_0,y-y_0)$ in coordinates $(x_0,y_0)$ (Fig. 7.6). The task is estimating target coordinates $(x_0,y_0)$ by mean of processing the observed image.

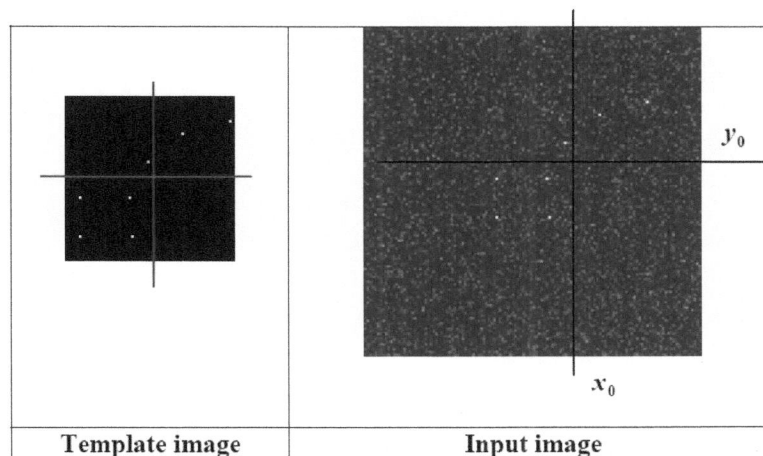

**Fig. 7.6.** Examples of template and input images for target location

There are many applications, in which observed images can be modeled as a sum

$$b(x,y)=a(x-x_0,y-y_0)+n(x,y) \tag{7.2.1}$$

of a template image $a(x-x_0,y-y_0)$ and a realzation $n(x,y)$ of additive Gaussian noise. In this ***additive gaussian noise (AGN-)*** model, the noise component is usually attributed to random interferences caused by the image sensor. Because of statistical nature of noise, the only option for estimating parameters $(x_0,y_0)$ is applying the statistical approach to optimal parameter estimation [5]. The best statistical estimates of parameters are the ***Maximum A-Posteriori Probability Estimation*** (***MAP-estimates***) and ***Maximum Likelihood*** (***ML-estimates***) Estimates. One can show ([6,7]) that, for the AGN-model, MAP and ML estimates $(\hat{x}_0, \hat{y}_0)$ of target coordinates are solutions of the equations, respectively:

$$\left\{ (\hat{x}_0, \hat{y}_0) \right\} = \underset{(x_0, y_o)}{\arg\max} \left\{ \int_{-\infty}^{\infty} \int_{-\infty}^{\infty} b(x,y) a(x-x_0, y-y_0) \, dx \, dy + N_0 \ln P(x_0, y_0) \right\}, \quad (7.2.2)$$

$$\left\{ (\hat{x}_0, \hat{y}_0) \right\} = \underset{(x_0, y_o)}{\arg\max} \left\{ \int_{-\infty}^{\infty} \int_{-\infty}^{\infty} b(x,y) a(x-x_0, y-y_0) \, dx \, dy \right\}, \quad (7.2.3)$$

where $N_0$ is spectral density of noise and $P(x_0, y_0)$ is probability density of target coordinates.

Thus, optimal ML-estimator should compute cross-correlation function between the target signal $a(x-x_0, y-y_0)$ and the observed signal $b(x,y)$ and take the coordinates of the maximum in the correlation pattern as the estimate. The optimal MAP-estimator also consists of a correlator and a decision-making device locating the maximum in the correlation pattern. The only difference between ML- and MAP-estimators is that, in the MAP-estimator, the correlation pattern is biased by the appropriately normalized logarithm of the object coordinates' a priori probability distribution.

The correlation operation for a mixture of signal and noise with a copy of the signal is often called ***matched filtering*** Correspondingly, the filter that implements this operation is called ***matched filter***. Schematic diagram of such an estimator is shown in Fig. 7.7.

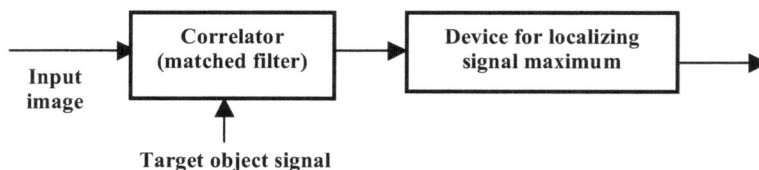

**Fig. 7.7.** Schematic diagram of the ML optimal device for target location in images

The correlation operation can be implemented in Fourier transform frequency domain by multiplication of the input signal spectrum by frequency response of the matched filter which, according to the properties of Fourier transform, is $\alpha^*(f_x, f_y)$, a complex conjugate to the target signal spectrum $\alpha(f_x, f_y)$. As it was mentioned above, the correlation, or matched filter type of the localization devices can be implemented by optical and holographic means.

In most of other applications, target objects should be located in images, which contain many other objects that may obscure the target object. For such images, the AGN model is not appropriate and it very frequently happens that cross-correlation between the target and other non-target objects exceeds the target object autocorrelation of which leads in the correlational target location system shown in Fig. 7.7 to frequent false detection. This phenomenon is illustrated in Fig. 7.8.

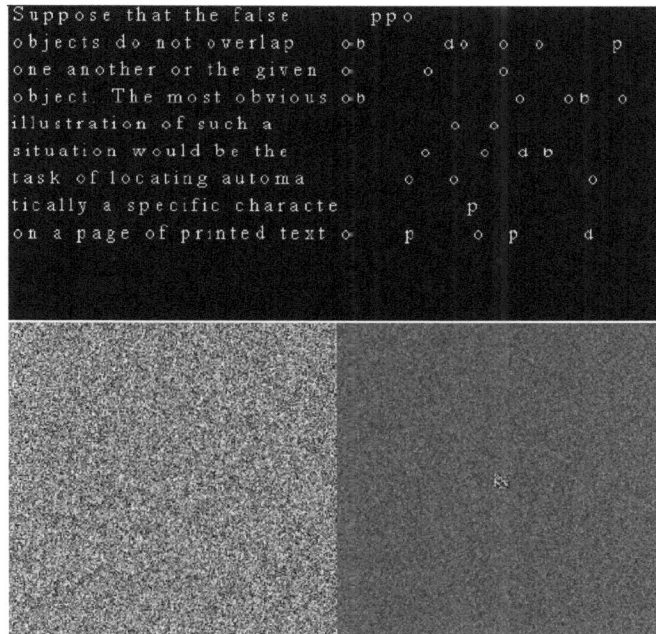

**Fig. 7.8.** Detection, by means of matched filtering, of a character with and without non-target characters in the area of search. Upper row, left image: noisy image of a printed text with standard deviation of additive noise 15 (within the signal range 0-255). Upper row, right image: results of detection of character "o" (right); one can see also quite a number of false detections of character "p" which are confused with character "o". Bottom row, left: a noisy image with a single character "o" in its center; standard deviation of the additive noise is 175. Highlighted center of the right image shows that the character is correctly localized by the matched filter.

In such applications, background non-target objects represent the main obstacle for reliable target localization, not the sensor's noise. In the rest of this section we show how can one the above correlation device optimal for the AGN model to minimize the danger of false identification of the target object with one of the non-target objects.

From a general point of view, this device is a special case of devices that consist of a linear filter followed by a unit for locating the highest peak of the signal at the filter output (Figure 7.9).

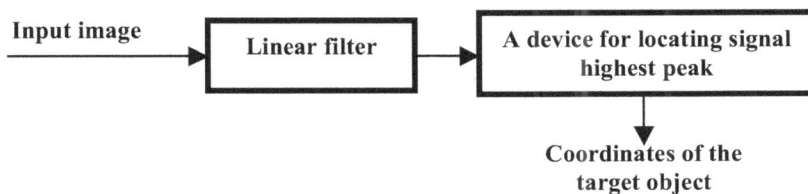

**Fig. 7.9.** Block diagram of a model of optical correlators for target detection and localization

If we restrict ourselves to such devices, which have above described electro-optical implementation, the linear filter should be optimized so as to minimize, on average over unknown target coordinates and image sensor's noise, the rate of false detections that occur in points, where signal at the filter exceeds the signal value at the point of the target location.

Let $b_0$ be the filter output signal value at the point $(x_0, y_0)$ of target location and $p(b_{bg}/(x_0, y_0))$ be the probability density of the filter output signal $b_{bg}(x, y)$ over the part of the image that does not contain the target object. Then the rate of false detections on average $AV_{Sens}AV_{TrgCoord}[.]$ the image sensor's noise (averaging operator $AV_{Sens}$) and over unknown target coordinates $(x_0, y_0)$ (averaging operator $AV_{TrgCoord}$) an be computed as

$$P_{fd} = AV_{Sens}AV_{TrgCoord}\left[\int_{b_0}^{\infty} p\left(b_{bg}/(x_0, y_0)\right) db_{bg}\right] =$$

$$\int_{AV_{Sens}(b_0)}^{\infty} AV_{Sens}AV_{TrgCoord}\left[p\left(b_{bg}/(x_0, y_0)\right)\right] db_{bg} \qquad (7.2.4)$$

This rate has to be minimized by the appropriate design of the linear filter.

For target signal $a(x - x_0, y - y_0)$ located at coordinates $(x_0, y_0)$, output of the filter with frequency response $H(f_x, f_y)$ at the point $(x_0, y_0)$ of the target location can be computed as

$$b_0 = \int_{-\infty}^{\infty}\int_{-\infty}^{\infty} \alpha(f_x, f_y) H(f_x, f_y) df_x \, df_y, \qquad (7.2.5)$$

where $\alpha(f_x, f_y)$ is Fourier spectrum of the target signal $a(x, y)$.

In order to analytically design the filter that minimizes the rate of false alarms, one needs a relationship between the filter frequency response and the probability density of the filter output signal given filter input signal. However, such a relationship is not known. It is known only that, by virtue of the central limit theorem, linear filtering tends to normalize distribution function of output signal and that the explicit dependence of $p(b_{bg}/(x_0, y_0))$ on the filter frequency response $H(f_x, f_y)$ can only be written for the second moment $m_2^2$ of the distribution using Parseval's relation for the Fourier transform:

$$m_2^2 = \int_{-\infty}^{\infty} b_{bg}^2 \, p\left(b_{bg} / (x_0, y_0)\right) db_{bg} = \int_{-\infty}^{\infty} b_{bg}^2 \left(x, y / x_0, y_0\right) dx \, dy =$$

$$\int_{-\infty}^{\infty} \int_{-\infty}^{\infty} \left|\beta_{bg}^{(0,0)}\left(f_x, f_y\right)\right|^2 \left|H\left(f_x, f_y\right)\right|^2 df_x \, df_y , \qquad (7.2.6)$$

where $\beta_{bg}^{(0,0)}\left(f_x, f_y\right)$ is Fourier spectrum of this background part of the image. Correspondingly, the second moment of the averaged probability distribution function $AV_{Sens} AV_{TrgCoord}\left[p\left(b_{bg} / (x_0, y_0)\right)\right]$ can be computed as

$$AV_{Sens} AV_{TrgCoord}\left[m_2^2\right] = \int_{-\infty}^{\infty} b_{bg}^2 \, AV_{Sens} AV_{TrgCoord}\left[h\left(b_{bg} / (x_0, y_0)\right)\right] db_{bg} =$$

$$AV_{Sens} AV_{TrgCoord}\left[\int_{-\infty}^{\infty} b_{bg}^2 \left(x, y / x_0, y_0\right) dx \, dy\right] =$$

$$AV_{Sens} AV_{TrgCoord}\left[\int_{-\infty}^{\infty} \int_{-\infty}^{\infty} \left|\beta_{bg}^{(0,0)}\left(f_x, f_y\right)\right|^2 \left|H\left(f_x, f_y\right)\right|^2 df_x \, df_y\right]. \qquad (7.2.7)$$

Therefore, for the analytical optimization of the localization device filter we will rely upon the Tchebyshev's inequality

$$\boldsymbol{Probability}\left(x \geq \overline{x} + b_0\right) \leq \sigma_x^2 / b_0^2 . \qquad (7.2.8)$$

that connects the probability that a random variable $x$ exceeds some threshold $\boldsymbol{b_0}$ and the variable's mean value $\overline{x}$ and standard deviation $\boldsymbol{\sigma_x}$.

Applying this relationship to Eq.(7.2.4), we obtain:

$$\overline{P}_a = \int_{\overline{b}_0}^{\infty} AV_{Sens} AV_{TrgCoord}\left[p\left(b_{bg}\right)\right] db_{bg} \leq AV_{bg} \frac{AV_{Sens} AV_{TrgCoord}\left[m_2^2 - \overline{b}^2\right]}{\left(AV_{Sens}\left[\overline{b}_0 - \overline{b}\right]\right)^2}, \qquad (7.2.9)$$

where $\overline{b}$ is mean value of the distribution function $p\left(b_{bg} / (x_0, y_0)\right)$.

By virtue of the properties of the Fourier Transform, the mean value of the background component of the image, $\overline{b}$, is determined by the filter frequency response $H(0,0)$ at the point $\left(f_x = 0, f_y = 0\right)$. Its value defines a constant bias of the signal at the filter output, which is irrelevant for the device that localizes the signal maximum. Therefore, one can choose $H(0,0) = 0$ and disregard $\overline{b}$ in Eq. (7.2.9). Then we can conclude that, in order to minimize the rate $\overline{P}_a$ of false detection errors, a filter should be found that maximizes the ratio of its response $\overline{b}_0$ to the target object to standard deviation $\left(m_2^2\right)^{1/2}$ of its response to the background image component

$$SClR = \frac{\overline{b}_o}{\left(m_2^2\right)^{1/2}} = \frac{\int\limits_{-\infty}^{\infty}\int\limits_{-\infty}^{\infty} AV_{Sens}\left[\alpha\left(f_x,f_y\right)\right]H\left(f_x,f_y\right)df_x\,df_y}{\left(\int\limits_{-\infty}^{\infty}\int\limits_{-\infty}^{\infty} AV_{Sens}\,AV_{TrgCoord}\left[\left|\alpha_{bg}^{0,0}\left(f_x,f_y\right)\right|^2\right]\left|H\left(f_x,f_y\right)\right|^2 df_x\,df_y\right)^{1/2}}$$

(7.2.10)

We will refer to this ratio as ***signal-to-clutter ratio (SClR)***.

From the Schwarz's inequality it follows that

$$\int\limits_{-\infty}^{\infty}\int\limits_{-\infty}^{\infty}\alpha\left(f_x,f_y\right)H\left(f_x,f_y\right)df_x\,df_y =$$

$$\int\limits_{-\infty}^{\infty}\int\limits_{-\infty}^{\infty}\frac{AV_{Sens}\left[\alpha\left(f_x,f_y\right)\right]}{\left(AV\left|\alpha_{bg}^{0,0}\left(f_x,f_y\right)\right|^2\right)^{1/2}}\left(AV\left|\alpha_{bg}^{0,0}\left(f_x,f_y\right)\right|^2\right)^{1/2}H\left(f_x,f_y\right)df_x\,df_y \le$$

$$\left\{\int\limits_{-\infty}^{\infty}\int\limits_{-\infty}^{\infty}\frac{\left|AV_{Sens}\left[\alpha\left(f_x,f_y\right)\right]\right|^2}{AV\left[\left|\alpha_{bg}^{0,0}\left(f_x,f_y\right)\right|^2\right]}df_x\,df_y\right\}^{1/2}\left\{\int\limits_{-\infty}^{\infty}\int\limits_{-\infty}^{\infty}\left|H\left(f_x,f_y\right)\right|^2\left(AV\left[\left|\alpha_{bg}^{0,0}\left(f_x,f_y\right)\right|^2\right]\right)df_x\,df_y\right\}^{1/2}.$$

(7.2.11)

with equality taking place when

$$H\left(f_x,f_y\right) = \frac{AV_{Sens}\left[\alpha^*\left(f_x,f_y\right)\right]}{AV\left[\left|\alpha_{bg}^{0,0}\left(f_x,f_y\right)\right|^2\right]},$$

(7.2.12)

where asterisk * denotes complex conjugation and $AV\left[.\right]$ denotes $AV_{Sens}\,AV_{TrgCoord}\left[.\right]$. Therefore signal-to-clutter ratio is

$$SClR \le \left\{\int\limits_{-\infty}^{\infty}\int\limits_{-\infty}^{\infty}\frac{\left|AV_{Sens}\left[\alpha\left(f_x,f_y\right)\right]\right|^2}{AV_{Sens}\,AV_{bg}\left[\left|\alpha_{bg}^{0,0}\left(f_x,f_y\right)\right|^2\right]}df_x\,df_y\right\}^{1/2}$$

(7.2.13)

reaching its maximum for the optimal filter defined by the equation:

$$H_{opt}\left(f_x,f_y\right) = \frac{AV_{Sens}\left[\alpha^*\left(f_x,f_y\right)\right]}{AV_{Sens}\,AV_{TrgCoord}\left[\left|\alpha_{bg}^{0,0}\left(f_x,f_y\right)\right|^2\right]},$$

(7.2.14)

This filter will be optimal for the particular observed image. Because its frequency response depends on the image background component the filter is adaptive. We will call this filter "***optimal adaptive correlator***".

To be implemented, the optimal adaptive correlator needs knowledge of power spectrum of the background component of the image. However coordinates of the target object are not known. Before the target object is located one cannot separate it from the background image component and, therefore, can not exactly determine the background component power spectrum and implement the exact optimal adaptive correlator. Yet, one can attempt to approximate the latter by means of an appropriate estimation of the background component power spectrum from the observed image.

There might be different approaches to substantiating spectrum estimation methods using additive and implant models for representation of the target and background objects within images.

In the additive model, input image $b(x,y)$ is regarded as an additive mixture of the target object $a(x-x_0, y-y_0)$ and image background component $a_{bg}(x,y)$:

$$b(x,y) = a(x-x_0, y-y_0) + a_{bg}(x,y), \tag{7.2.15}$$

where $(x_0, y_0)$ are unknown coordinates of the target object. Therefore power spectrum of the image background component averaged over the set of possible target locations can be estimated as

$$AV_{TrgCoord}\left[\left|\alpha_{bg}^{(0,0)}(f_x,f_y)\right|^2\right] = \left|\beta(f_x,f_y)\right|^2 + \left|\alpha(f_x,f_y)\right|^2 +$$

$$\beta^*(f_x,f_y)\alpha(f_x,f_y)AV_{TrgCoord}\left[\exp\left[i2\pi(f_x x_0 + f_y y_0)\right]\right] +$$

$$\beta(f_x,f_y)\alpha^*(f_x,f_y)AV_{TrgCoord}\left[\exp\left[-i2\pi(f_x x_0 + f_y y_0)\right]\right], \tag{7.2.16}$$

Functions $\quad AV_{TrgCoord}\left\{\exp\left[i2\pi(f_x x_0 + f_y y_0)\right]\right\}\quad$ and $AV_{TrgCoord}\left\{\exp\left[-i2\pi(f_x x_0 + f_y y_0)\right]\right\}$ are Fourier transforms of the distribution density $p(x_0, y_0)$ of the target coordinates:

$$AV_{TrgCoord}\left\{\exp\left[\pm i2\pi(f_x x_0 + f_y y_0)\right]\right\} = \iint_{X Y} p(x_0, y_0)\exp\left[\pm i2\pi(f_x x_0 + f_y y_0)\right]dx_0\, dy_0 \tag{7.2.17}$$

In the assumption that the target object coordinates are uniformly distributed within the input image area, these functions are sharp peak functions with maximum at $f_x = f_y = 0$ and negligible values for all other frequencies. We agreed above that point $f_x = f_y = 0$ is not relevant for the filter design because filter frequency response in this point defines filter output signal constant bias. Therefore, for the additive model of the target object and image background component, averaged power spectrum of the background component may be estimated as:

$$AV_{TrgCoor}\left[\left|\alpha_{bg}^{(0,0)}\left(f_x,f_y\right)\right|^2\right] \cong \left|\beta\left(f_x,f_y\right)\right|^2 + \left|\alpha\left(f_x,f_y\right)\right|^2 ; f_x,f_y \neq 0 .(7.2.18)$$

for all relevant points $f_x,f_y \neq 0$ in frequency domain.

The implant model assumes that

$$a_{bg}\left(x,y\right) = b\left(x,y\right)w\left(x-x_0,y-y_0\right), \tag{7.2.19}$$

where $w\left(x-x_0,y-y_0\right)$ is a target object outlining window function:

$$w\left(x,y\right) = \begin{cases} 0 & \text{within the target object} \\ 1, & \text{elsewhere} \end{cases} . \tag{7.2.20}$$

In a similar way as it was done for the additive model and in the same assumption of uniform distribution of target coordinates over the image area, one can show that in this case power spectrum of the background image component averaged over all possible target coordinates can be estimated as

$$AV_{TrgCoor}\left[\left|\alpha_{bg}^{0,0}\left(f_x,f_y\right)\right|^2\right] \cong \iint\left|W\left(p_x,p_y\right)\right|^2\left|\beta\left(f_x-p_x,f_y-p_y\right)\right|^2 dp_x\,dp_y, \tag{7.2.21}$$

where $W\left(f_x,f_y\right)$ is Fourier transform of the window function $w\left(x,y\right)$. Noteworthy that this method of estimating background component power spectrum resembles the traditional methods of signal spectra estimation that assume convolving power spectrum of the signal with a certain smoothing spectral window function $\bar{W}\left(p_x,p_y\right)$ (see, for example, [8]).

Both models imply that, as a zero order approximation, the row input image power spectrum can be used as an estimate of the background component power spectrum:

$$AV_{TrgCoor}\left[\left|\alpha_{bg}^{0,0}\left(f_x,f_y\right)\right|^2\right] \cong \left|\beta\left(f_x,f_y\right)\right|^2 . \tag{7.2.22}$$

This approximation is justified by the natural assumption that the target object size is substantially smaller than the size of the input image and that its contribution to the entire image power spectrum on most frequencies is negligibly small with respect to that of the background image component.

As for the averaging over image sensor's noise, one can show that, in the assumption of additive signal-independent zero mean sensor noise, this averaging will result in adding to the above estimates (7.2.18), (7.2.21) and (7.2.22 variance $\sigma_n^2$ of the noise. The same averaging over the image sensor's noise required, according to Eq. 7.2.14, for the target spectrum does not change it under the above assumption of additive signal-independent zero mean noise. In this way we arrive at the following three modifications of optimal adaptive correlators:

$$H_{opt}\left(f_x,f_y\right) = \frac{\alpha^*\left(f_x,f_y\right)}{\left|\beta\left(f_x,f_y\right)\right|^2 + \left|\alpha\left(f_x,f_y\right)\right|^2 + \sigma_n^2} \qquad (7.2.23)$$

$$H_{opt}\left(f_x,f_y\right) = \frac{\alpha^*\left(f_x,f_y\right)}{\displaystyle\int_{-\infty}^{\infty}\int_{-\infty}^{\infty}\left|W\left(p_x,p_y\right)\right|^2\left|\beta\left(f_x-p_x,f_y-p_y\right)\right|^2 dp_x\,dp_y + \sigma_n^2} \qquad (7.2.24)$$

$$H_{opt}\left(f_x,f_y\right) = \frac{\alpha^*\left(f_x,f_y\right)}{\left|\beta\left(f_x,f_y\right)\right|^2 + \sigma_n^2} \qquad (7.2.25)$$

## 7.3 The variety of optical correlators and comparison

Since invention of the optical correlator-matched filter by VanderLugt ([1]), a variety of optical correlators have been suggested and studied:

- ***Matched filter (MF) correlator*** ([1])

$$OUTPUT = FT\left\{FT\left(INPUT\right)\bullet\left(FT\left(TRTobj\right)\right)^*\right\} \qquad (7.3.1)$$

- ***Phase-only filter (POF) correlator*** ([9])

$$OUTPUT = \mathbf{FT}\left\{\mathbf{FT}\left(INPUT\right)\bullet\frac{\left(\mathbf{FT}\left(TRTobj\right)\right)^*}{\left|\mathbf{FT}\left(TRTobj\right)\right|}\right\} \qquad (7.3.2)$$

- ***Phase-only (PO) correlator*** ([10])

$$OUTPUT = \mathbf{FT}\left\{\frac{\mathbf{FT}\left(INPUT\right)}{\left|\mathbf{FT}\left(INPUT\right)\right|}\bullet\frac{\left(\mathbf{FT}\left(TRTobj\right)\right)^*}{\left|\mathbf{FT}\left(TRTobj\right)\right|}\right\} \qquad (7.3.3)$$

- ***Optimal adaptive correlator*** (OAC) ([6])

$$OUTPUT = \mathbf{FT}\left\{ \frac{\mathbf{FT}(INPUT) \bullet \left(\mathbf{FT}(TRTobj)\right)^*}{\overline{\left|\mathbf{FT}(INPUT)\right|^2}} \right\} =$$

$$\mathbf{FT}\left\{ \frac{\mathbf{FT}(INPUT)}{\left[\overline{\left|\mathbf{FT}(INPUT)\right|^2}\right]^{1/2}} \frac{\left(\mathbf{FT}(TRTobj)\right)^*}{\left[\overline{\left|\mathbf{FT}(INPUT)\right|^2}\right]^{1/2}} \right\} \qquad (7.3.4)$$

- *(-k)-th law nonlinear correlator* ([11])

$$OUTPUT = \mathbf{FT}\left\{ \frac{\mathbf{FT}(INPUT) \bullet \left(\mathbf{FT}(TRTobj)\right)^*}{\left(\overline{\left|\mathbf{FT}(INPUT)\right|^2}\right)^k} \right\} \qquad (7.3.5)$$

- *Joint Transform Correlator* ([12])

$$\mathbf{OUTPUT} = \mathbf{FT}\left\{ \left|\mathbf{FT}(\mathbf{INPUT} + \mathbf{TRTobj})\right|^2 \right\} \qquad (7.3.6)$$

- Binarized versions of the correlators: amplitude or phase components of the correlator filter or both are binarized

In these formulas $\mathbf{FT}(.)$ denotes Fourier Transform operator, $INPUT$ , $OUTPUT$ and $TRTobj$ denote input image, output image and reference object image, correspondingly.

Matched filter correlators, phase-only filter correlators and phase-only correlators can be optically implemented in optical setups shown in Figs. 7.4 and 7.5 with spatial filters recorded on spatial light modulators in Fourier planes of these setups. As these spatial filters, computer generated holograms can be used.

In the case of the phase-only filter correlator, only the phase component of the matched filter is recorded. In the case of the phase-only correlator, also only the phase component of the matched filter is recorded but this filter is used to modulate spectrum of the input image, in which amplitude component is forcibly set to constant. For this, especial arrangements are required, which we will not discuss here. The reader may want to refer to Ref. 10.

Optical implementation of the optimal adaptive correlators requires using a non-linear signal transformation in the Fourier plane of the correlators. One of the option is inserting in Fourier planes of setups of Figs. 7.4 and 5 a nonlinear medium with a transfer function

$$\textbf{\textit{Output Amplitude}} = \left(\textbf{\textit{Input amplitude}}\right)^{-k} \qquad (7.3.7)$$

We will refer to this type on nonlinearity as *(-k)-th law nonlinearity*. A plate with this medium can be placed at Fourier planes of the systems slightly out of focus in order to image spectrum smoothing before its non-linear transformation as it is recommended by Eq. 7.2.21 for better estimation of the background image component power spectrum. Block diagrams of the nonlinear optical correlator with (-*k*)-th law nonlinearity built on the base of the lens system of Fig. 7.4 is shown in Figure 7.10. As the intensity of light incoming to the nonlinear medium is proportional to the intensity of input image power spectrum ($\left|\hat{E}\left(f_x, f_y\right)\right|^2$ in Eq. 7.2.22), optimal value for the parameter **k** of the nonlinearity is $k = 1$. Simulation experiments reported in Ref. 11 show that slight deviations from this optimal value are tolerable.

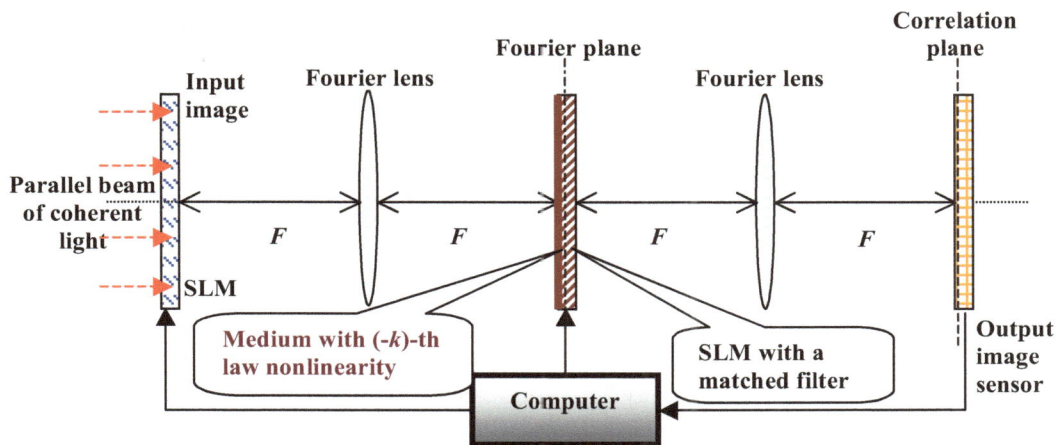

**Fig. 7.10.** Lens based nonlinear electro-optical correlator. **F** – focal distance of Fourier lenses.

Similar modification of the parabolic mirror system is also possible. In this case, requirements to the dynamic range of the nonlinear medium are eased because light propagates through the nonlinear medium twice, before coming to the reflective filter and after the reflection.

Described modifications of optical correlators differ in their capability of discriminating target objects from non-target objects and clutter background in images. In Figure 7.11, results of experimental comparison of the discrimination capability of above described optical correlators reported in Ref. 11 are presented. In the experiments, carried out using 16 test air photographs of 128x128 pixels presented in Fig. 7.12 and a test target image in a form of a circle of 5 pixels in diameter embedded into test images, signal-to-clutter ratio was measured for each of the correlator. In addition to all described correlators, the ideal optimal adaptive correlator was tested as well, built for exact background component of images, which was known in the simulation experiments.

Results presented in Fig. 7.9 show that optimal adaptive correlator considerably outperforms other types of correlators in terms of signal-to-clutter ratio they provide though still there is a two to five fold gap between its performance and that of the exact optimal adaptive correlator. This gap motivates search for better methods of estimating background component of images.

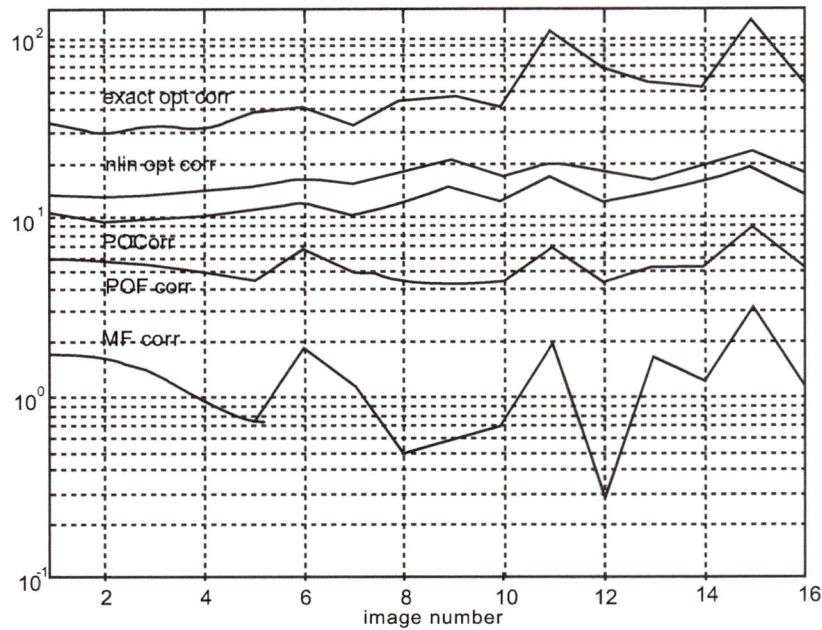

**Fig. 7.11.** Comparison, in terms of the signal-to-clutter ratio (*SCIR*), of the discrimination capabilities of the exact Optimal Adaptive Correlator (exact opt corr.); optimal adaptive correlator (nlin. opt. corr.) with background image component power spectrum estimation by means of blurring observed image power spectrum according Eq. 7.2.21; phase-only correlator (POCorr), phase-only filter correlator (POF corr), and matched filter correlator (MF corr) for a set of 16 test images of 128x128 pixels and a circular target object of 5 pixels in diameter embedded in images.

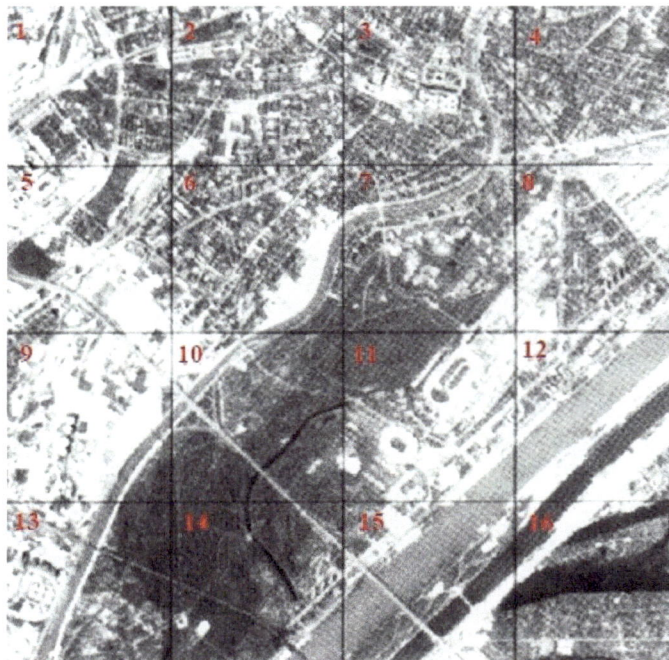

**Fig. 7.12.** Set of 16 test images of 128x128 pixels

## 7.4   Joint Transform Correlator

Lens and parabolic mirror correlators require pre-fabricated filter placed in their Fourier plane or a computer-generated hologram synthesized and recorded by computer. Joint Transform correlator is an electro-optical correlator that does not require a pre-fabricated filter. Schematic diagram of the Joint Transform Correlator (JTC) is presented on Fig. 7.13. JTC works as following. Let, as before, $b(x,y)$ be an input image and $a(x,y)$ be a template image that should be located in the input image. In the Joint Transform Correlator, the template image is placed aside to the input image in the same front focal plane of the first Fourier lens. The lens performs Fourier transform of the sum $b(x,y)+a(x,y)$ of these images and forms in its rear focal plane their joint spectrum $\gamma(f_x,f_y)=\beta(f_x,f_y)+\alpha(f_x,f_y)$. A photosensitive sensor installed in the lens rear focal plane generates an electrical signal proportional to a certain, in general nonlinear, function $g\left(\left|\gamma(f_x,f_y)\right|^2\right)$ of the joint spectrum intensity:

$$g\left(\left|\gamma(f_x,f_y)\right|^2\right)=g\left(\left|\beta(f_x,f_y)+\alpha(f_x,f_y)\right|^2\right)=$$
$$g\left(\left|\beta(f_x,f_y)\right|^2+\left|\alpha(f_x,f_y)\right|^2+\beta^*(f_x,f_y)\alpha(f_x,f_y)+\beta(f_x,f_y)\alpha^*(f_x,f_y)\right) \quad (7.4.1)$$

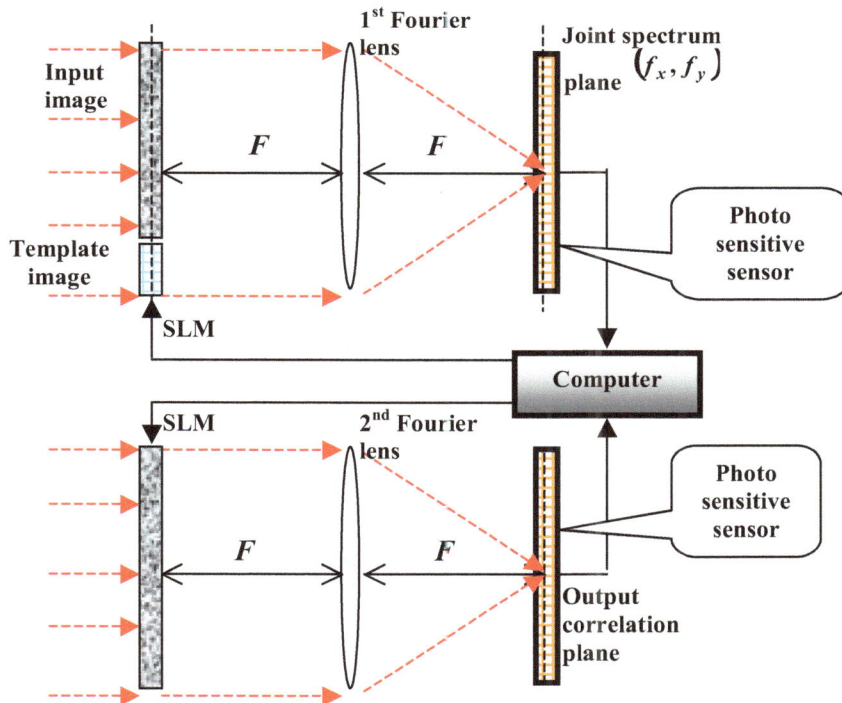

**Fig. 7.13.** Joint Transform correlator. $F$ – focal distance of the Fourier lenses.

As one can see in Eq. 7.3.8, the last term in the argument of the function $g(\cdot)$ is the product of the input image spectrum $\beta\left(f_x, f_y\right)$ and complex conjugate $\alpha^*\left(f_x, f_y\right)$ to the template image spectrum, which corresponds to matched filtering in transform domain.

Signal $g\left(\left|\gamma\left(f_x, f_y\right)\right|^2\right)$ can be, in principle, further modified by the computer or directly used to record in on the second SLM installed in frontal Fourier plane of the second Fourier lens of the JTC. This performs its Fourier transform and forms a correlation image in its rear Fourier plane, where it is sensed by another photosensitive sensor and put in the computer for further processing and analysis.

Input image and template image can, in principle, be optically recorded on the input SLM. Recording them and, especially, the template image, from computer is more advantageous, as it enables rapid change of templates when detection of different modifications of target images, such as scaled and rotated, is required.

If the function $g(\cdot)$ is a linear function, as it is commonly assumed, Joint Transform Correlator can be used as a matched filter correlator. However, with an appropriate selection of the nonlinear function $g(\cdot)$, Joint Transform Correlator can approximate the optimal adaptive correlator.

Consider frequency response (Eq. 7.2.23) of the optimal adaptive correlator obtained for the additive model of evaluating power spectrum of the background component of images and neglect in it the variance of the additive sensor's noise:

$$H_{opt}\left(f_x, f_y\right) \cong \frac{\alpha^*\left(f_x, f_y\right)}{\left|\beta\left(f_x, f_y\right)\right|^2 + \left|\alpha\left(f_x, f_y\right)\right|^2}, \qquad (7.4.2)$$

where $\beta\left(f_x, f_y\right)$ and $\alpha\left(f_x, f_y\right)$ are Fourier spectra of the image and the target object, correspondingly, and let us show that, with a logarithmic nonlinear transformation of the joint spectrum, the JTC approximates this filter. With the logarithmic nonlinearity, the transformed joint power spectrum $\left|\gamma\left(f_x, f_y\right)\right|^2$ at the output of this nonlinear device can be written as

$$Log\left|\gamma\left(f_x, f_y\right)\right|^2 = \log\left\{\left|\beta\left(f_x, f_y\right) + \alpha\left(f_x, f_y\right)\right|^2\right\} =$$
$$\log\left\{\left|\beta\left(f_x, f_y\right)\right|^2 + \left|\alpha\left(f_x, f_y\right)\right|^2 + \beta^*\left(f_x, f_y\right)\alpha\left(f_x, f_y\right) + \beta\left(f_x, f_y\right)\alpha^*\left(f_x, f_y\right)\right\} \quad (7.4.3)$$

Since the size of the reference object is usually much smaller than the size of the input image, one can assume that, for the majority of the spectral components,

$$\left|\beta\left(f_x, f_y\right)\right|^2 + \left|\alpha\left(f_x, f_y\right)\right|^2 \gg \beta^*\left(f_x, f_y\right)\alpha\left(f_x, f_y\right) + \beta\left(f_x, f_y\right)\alpha^*\left(f_x, f_y\right) \quad (7.4.4)$$

With this assumption, $Log\left|\gamma\left(f_x,f_y\right)\right|^2$ is approximately equal to

$$LogJPS\left(f_x,f_y\right) \cong \log\left\{\left|\beta\left(f_x,f_y\right)\right|^2 + \left|\alpha\left(f_x,f_y\right)\right|^2\right\} + \frac{\beta^*\left(f_x,f_y\right)\alpha\left(f_x,f_y\right)}{\left|\beta\left(f_x,f_y\right)\right|^2 + \left|\alpha\left(f_x,f_y\right)\right|} +$$

$$\frac{\alpha^*\left(f_x,f_y\right)}{\left|\beta\left(f_x,f_y\right)\right|^2 + \left|\alpha\left(f_x,f_y\right)\right|}\beta^*\left(f_x,f_y\right) \qquad (7.4.5)$$

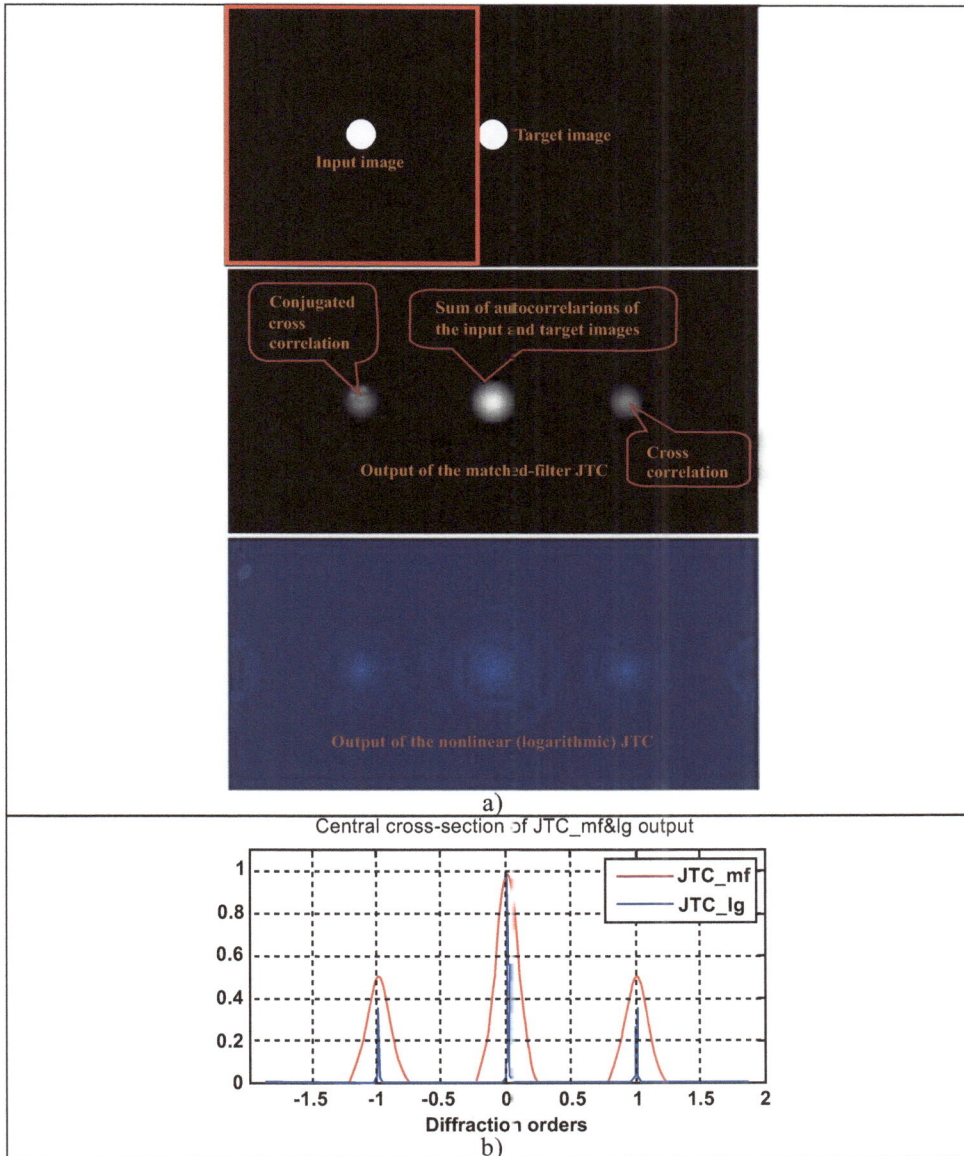

**Fig. 7.14.** Arrangement of input image and target object at the input of JTC and resulting correlation outputs of conventional linear (matched-filter) and nonlinear (logarithmic) JTCs (a); and central cross sections of the matched-filter and nonlinear JTCs outputs for the test input and target images shown in a) (b).

When a JTC configuration is utilized, the two last terms displaying the correlation function and its conjugate are readily separated from each other and from the first term that produces the zero order diffraction term (see the arrangement in Fig. 7.14). The last term of expression (7.4.5) exactly reproduces filtering described by Eq. 7.4.2. Therefore, one can conclude that the joint transform correlator with a logarithmic non-linearity can be regarded as an approximation to the OAC and therefore promises an improved discrimination capability.

The logarithmic nonlinearity can be implemented in JTCs by computer processing of the joint spectrum or directly in the photosensitive sensor on an analog level. Optical sensors with a logarithmic sensitivity functions are now becoming commercially available.

## References

1.   A. VanderLugt. Optical Signal Processing.Wiley, New York, 1992
2.   F. T. S. Yu. Optical Information Processing. Wiley, New York, 1983
3.   F. T. S. Yu and S. Jutamulia, Eds., Optical Pattern Recognition. Cambridge University Press, Cambridge, 1998
4.   A. B. VanderLugt, "Signal detection by complex matched spatial filtering," IEEE Trans. Inf. Theory IT-10, 139-145 (1964).
5.   H. L. Van-Trees, Detection, Estimation and Modulation Theory, Wiley, 1968
6.   L. P. Yaroslavsky, "The Theory of Optimal Methods for Localization of Objects in Pictures," *Progress in Optics Series,* Vol. 32 (Elsevier, Amsterdam, 1993), pp. 147-201.
7.   L.P. Yaroslavsky, Digital Holography and Digital Image Processing, Kluwer Academic Publisher, Boston, 2004
8.   R.B. Blackman and J. W. Tukey, The Measurement of Power Spectra from the Point of View of Communications Engineering, N.Y. Dover, 1959
9.   J. L. Horner and P. D. Gianino, "Phase-only matched filtering," Appl. Opt. 23, 812-816 (1984).
10.  T. Szoplik and K. Chalasinska-Macukow, ''Towards nonlinear optical processing,'' in International Trends in Optics, J. W. Goodman, ed. 1Academic, Boston, San Diego, New York, 1991, pp. 451–464.
11.  L. P. Yaroslavsky, "Optical correlators with $(-k)$-th law non-linearity: Optimal and suboptimal solutions," Appl. Opt. 34, 3924-3932 (1995).
12.  F. T. S. Yu and X. J. Lu, "A real-time programmable joint transform correlator," Opt. Commun. 52, 10-16 (1984).
13.  L. Yaroslavsky, E. Marom, Nonlinearity Optimization in Nonlinear Joint Transform Correlators, Applied Optics, vol. 36, No. 20, 4816-4822 (1997). (See also in: Selected Papers on Optical Pattern Recognition Using Joint Transform Correlation, M. S. Alam, ed., Milestone Series, v. MS 157, SPIE Optical Engineering Press, 1999

# 8     Computer-Generated Holograms and 3D Visual Communication

**Abstract:** This chapter is devoted to synthesis and application of computer-generated display holograms for 3D visualization and communication. It introduces a computer generated hologram based 3-D communication paradigm, discusses properties and limitations of human visual system relevant to 3D visualization and describes several specific methods for synthesis of computer-generated hologram capable of reproducing 3D sensation and possible approaches to designing 3D holographic displays

One of the motivations of computer-generated holography was the desire to ultimately solve the problem of 3D visual communication. The idea of producing artificial optical elements capable when illuminated of creating virtual images can be, perhaps, traced back to middle-ages Chinese mirrors (Fig. 8.1). The inventions in holography ([1, 2]) by E. Leith and Yu. Denisyuk (Figs. 8.2 and 3) were primarily motivated by the desire to create efficient means for visualizing 3D images.

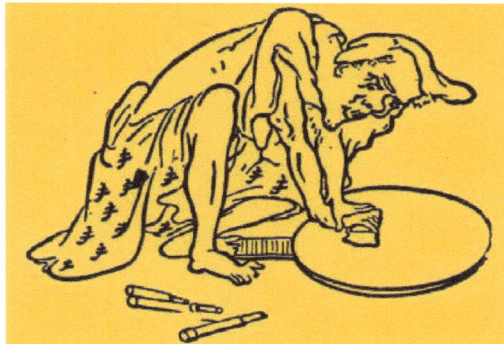

Engraving Chinese mirror

Image that appears when the mirror is illuminated

**Fig. 8.1**. Chinese mirrors – a middle ages prototype of computer-generated holography

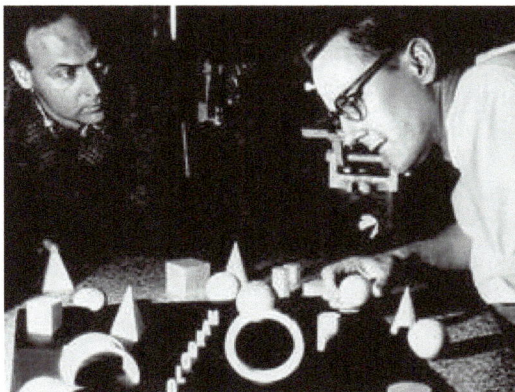

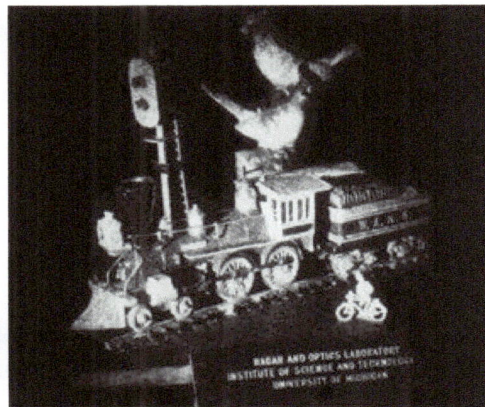

**Fig. 8.2.** Emmett Leith and Juris Upatnieks (left and right on the left image) at the University of Michigan producing their "Train and Bird " hologram, the first hologram ever made with a laser using the off-axis technique. Right image shows the reconstructed image.

**Fig. 8.3.** Yury N. Denisyuk with his holographic portrait

There are no doubts that holographic imaging is an ultimate solution for 3-D visualization. This is the only method that is capable of reproducing, in the most natural viewing conditions, 3-D images that have all the visual properties of the original objects including full parallax, and are visually separated from the display device. 3-D visual communication and display can be achieved through generating, at the viewer side, of holograms out of data that contain all relevant information regarding the scene to be viewed. Digital computers are ideal means for converting data on 3-D scenes into optical holograms for visual perception ([3, 4]).

In this chapter, we formulate a 3D communication paradigm based on computer-generated display holograms, review properties of human visual system relevant to 3D visualization and limitations of binocular vision, one of the most important mechanisms of 3D vision, and describe several practical methods of synthesis computer-generated display holograms.

## 8.1  Computer generated display holograms based 3D communication paradigm

The core of the 3D digital holographic visual communication paradigm is the understanding that, for generating synthetic holograms at the viewer side, one does not need to produce, at the scene side, the hologram of the scene and to transmit it to the viewer's site. Neither does one need to necessarily imitate, at the viewer site, the full optical holograms of the scene. What one does need is to collect, at the scene side, a set of data that will be sufficient to generate, at the viewer site, a synthetic hologram of the scene for viewing.

What features of optical holograms computer-generated display holograms for 3D visualization should have? The major requirement to computer-generated display holograms is that they should involve all relevant visual mechanisms of 3D perception and provide natural viewing conditions for the human visual system including, in particular, separation of reconstructed images from the display device.

A crucial issue in transmitting data needed for the synthesis, at the viewer site, of display holograms is the volume of data to be collected and transmitted and the computational complexity of the hologram synthesis. The upper bound of the amount of data needed to be collected at the scene side and transmitted to the viewer site is, in principle, the full volumetric description of the scene geometry and optical properties.

However, a realistic estimation of the amount of data needed for generating a display hologram of the scene is by orders of magnitude lower then the upper bound due to the limitations of the human visual system. This also has a direct impact on the computational complexity of the hologram synthesis.

## 8.2 Properties of human visual system relevant to holographic 3D visualization

The human visual system has quite a number of natural limitations that must be taken into account in the synthesis of computer-generated display holograms:

- Human vision works in incoherent light;
- At every particular moment, each of two eyes perceives only a small fraction of the incoming wave front limited by the size of the pupil (about 3x3 mm);
- 3D perception is achieved through several mechanisms complementing each other:
  - Eye accommodation;
  - Eye convergence, or the inward rotation of the eyes to converge on objects as they move closer to the observer;
  - Binocular disparity, or the difference in the images projected on the left and right eye retinas in the viewing of a 3D scene. Binocular disparity is the salient depth cue used by the visual system to produce the sensation of depth, or stereopsis.
  - Shading, shadowing and play of highlights on diffuse surfaces that do not have edges or textures capable of producing perception of binocular disparity;
  - Interposition, or occluding, hiding or overlapping one object by another;
  - Retinal image size;
  - Linear perspective;
  - Aerial perspective;
  - Motion parallax, which provides different views of a scene in response to movement of the scene or the viewer.

## 8.3 Binocular vision and its limitations

### 8.3.1. Redundancy of stereoscopic images: a qualitative evaluation

It is well known that stereoscopic images are very redundant. One can make an estimation of the redundancy of stereoscopic images using the following rationale. From the informational point of view, two images of the same scene that form a stereo pair are equivalent to one of the images and a depth map of the scene. Indeed, from two images of the stereo pair, one can build the depth map, and, vice-versa, one can build a stereo pair from one image and the depth map. Therefore, the increase of the signal volume the second image of the stereo pair adds to that of one image is equal to the signal volume that corresponds to the depth map. The number of depth gradations resolved by vision is of the same order of magnitude as the number of resolved image gray levels. Therefore, signal volume increment due to the depth map is basically determined by the number of depth map independent samples.

Every sample of the depth map can be found by localizing corresponding fragments of two images of the stereo pair and measuring parallax between them. All technical devices that measure depth from stereo work in this way, and it is only natural to assume the same mechanism for stereoscopic vision. The number of independent measurements of the depth map is obviously the ratio of the image area to the minimal area of the fragments of one image that can be reliably localized on another image. It is also obvious, that it is, generally, not possible to reliably localize one pixel of one image on another image. For reliable localization, image fragments should contain several pixels. Therefore, the number of independent samples of the depth map will be, correspondingly, several times lower than the number of image pixels, and the increment of the signal volume that corresponds to the depth map will be several times lower than the signal volume of one image. For instance, if the reliable size of the localized fragment is 2x2 pixels, it will be four times lower, for 3x3 fragments, it will be 9 times lower, and so on. Practical experience tells that, for reliable localization of fragments of one image on another image, the fragment size should usually exceed the area of 8x8 to 10x10 pixels.

Therefore one can hypothesize that the signal volume increment associated with the depth map may amount to percents or even fractions of a percent of the signal volume of one image and that this limitation of the resolution of depth maps acquired from stereo images is true for any device for extraction of depth information from stereo images, including human stereoscopic vision.

### 8.3.2.   *Tolerance of 3D visual perception to image blur: quantitative data*

The redundancy of stereoscopic images is a direct indication of limitations of binocular vision.  One of manifestations of these limitations is tolerance of human visual system to blur of one of two stereo images.

Perhaps one of the first observations of the phenomenon of the tolerance of stereoscopic vision to blur of one of stereo images was made by B. Julesz [5], who however did not mention any quantitative measures of this phenomenon.  Ref. [6]) provides some quantitative data. They are summarized in Fig. 8. 4 that shows root mean squared error of parallax measurements on a training stereo air photograph analyzed by a professional human operator using stereo-comparator when one of the images was decimated with different decimation factors and then interpolated back to the initial size. The data plotted in Fig. 8.4 were obtained by averaging of the parallax measuring errors over 31 randomly selected image fragments. Decimation and interpolation were carried out in a computer and the operator was working with computer-generated images of a good photographic quality produced from scanned original photos. Decimation factor zero on the bar graph corresponds to data obtained for original not scanned photos (not computer printouts).

One can see from these data that 4x- and even 5x-decimation/interpolation of one of two images of a stereo pair do not dramatically increase the measurement error. With the increase of the decimation order, RMS error grows according to a parabolic law. These experiments showed also that, after 7x-decimation/interpolations, localization failures appear, and the probability of failures grows with further increase of the decimation factor very rapidly. All this is in a good correspondence with the theory of localization accuracy of image correlators ([7]), although the data were obtained for a human operator.

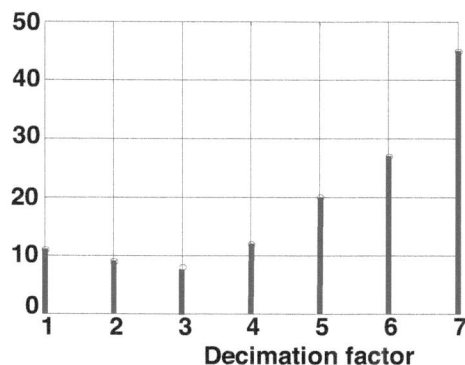

**Fig. 8.4.** Root mean squared error of parallax measurements (in units of the stereo-comparator control handle scale) as a function of the decimation factor as measured on 31 randomly selected fragments of a training stereo air photograph analyzed by a professional human operator

In an extended series of experiments on human stereo vision tolerance to image blur reported in [8] measured were stereopsis threshold and parallax measurement accuracy as functions of the degree of blur of one of two stereo images. As test images, As test images, the following images were used:
- Grayscale random dots images (such as one shown in Fig. 8. 5(a))
- Grayscale random patches images (Fig. 8.5 (b))
- Grayscale texture images (Figs. 8.5 (c), 8.5 (d))
- Color random patches images (Figs. 8.5 (e), 8.5 (f))
- Real-life stereoscopic images (see an example in Figs. 8.5 (g), 8.5 (h), right and left, respectively, for crossed eyes viewing)

For each of the test images except for real-life ones, its artificial stereo pair was artificially generated using, as a depth map, blobs of square or round shapes of different size on a uniform background (Figs. 8.6 (a), and (b)). Fig. 8.7 shows examples of stereo images generated in this way.

For each particular experiment, blobs defining the depth map were randomly placed within the image area so as to exclude the viewer's adaptation. In the experiments on the stereopsis threshold, the tested values of the depth map scale parameter that determines image parallax were randomly reshuffled for the same purpose. In addition, before displaying each next image, the display was kept blank for several seconds to secure the absence of a cross talk between individual experiments.

All experiments were carried out with several viewers. For each viewer and for each of the measured parameters, statistics were accumulated over several tens of realizations of random depth map impulse positions. The obtained data were statistically averaged using arithmetic mean and median averaging and standard deviation of the measured data was determined. In addition, time delay between the prompt to viewer to respond and the viewer's response was measured in all experiments in order to obtain supplementary data to the basic ones

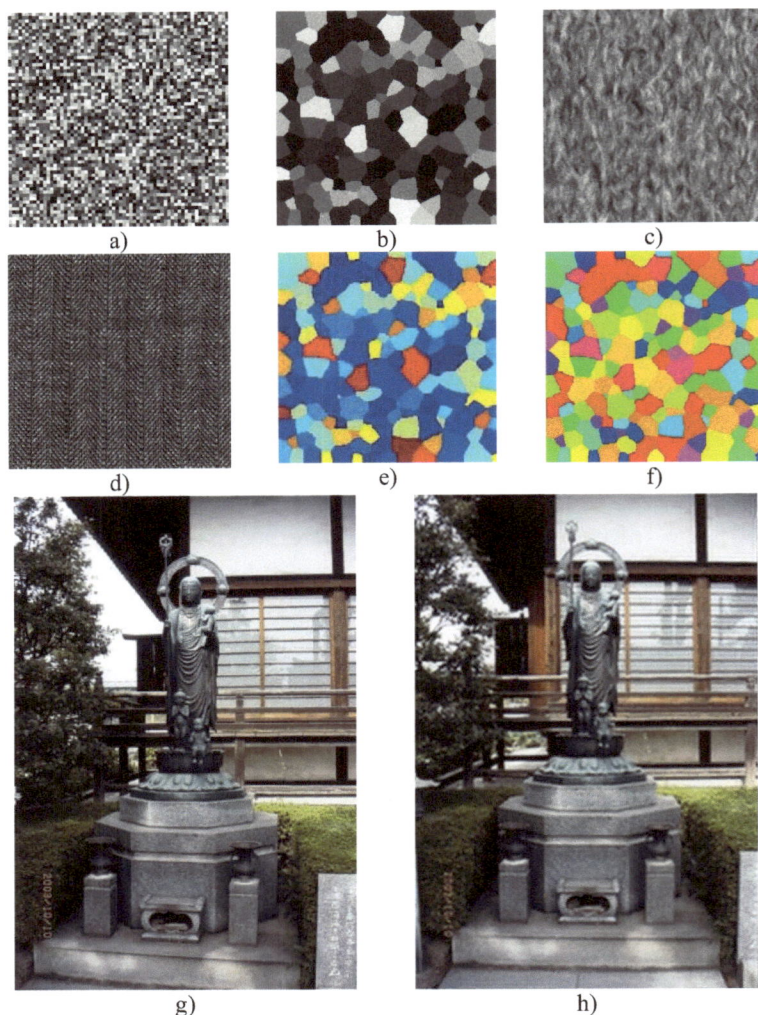

**Fig. 8.5.** Test images used in the experiments. Image h) exemplifies low pass filtering initial image to 1/4 of the image base band. Note that when images g) and h) are fused using stereoscope or by crossed eyes, a 3-D sharp image can be seen (it is recommended to magnify images for viewing)

In the experiments on the stereopsis threshold, synthetic stereo images with different parallax were displayed for viewers who were requested to detect a 3D target by indicating its position using computer generated cross cursor. Computer program registered the correct detection if the cursor was placed within the target object. For each degree of blur, from no blur to maximal blur, the target in form of a circle was randomly placed within the image and its parallax was changed from low to higher values until viewer detects the 3D target after which a new session with another blur started. Both random dot and texture test images were used in the experiments.

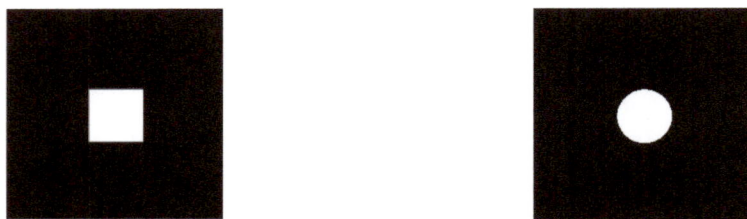

**Fig. 8.6.** Depth maps used in the experiments

a) Right eye image

b) Left eye image, low pass filtered to ¼ of the base band

c)

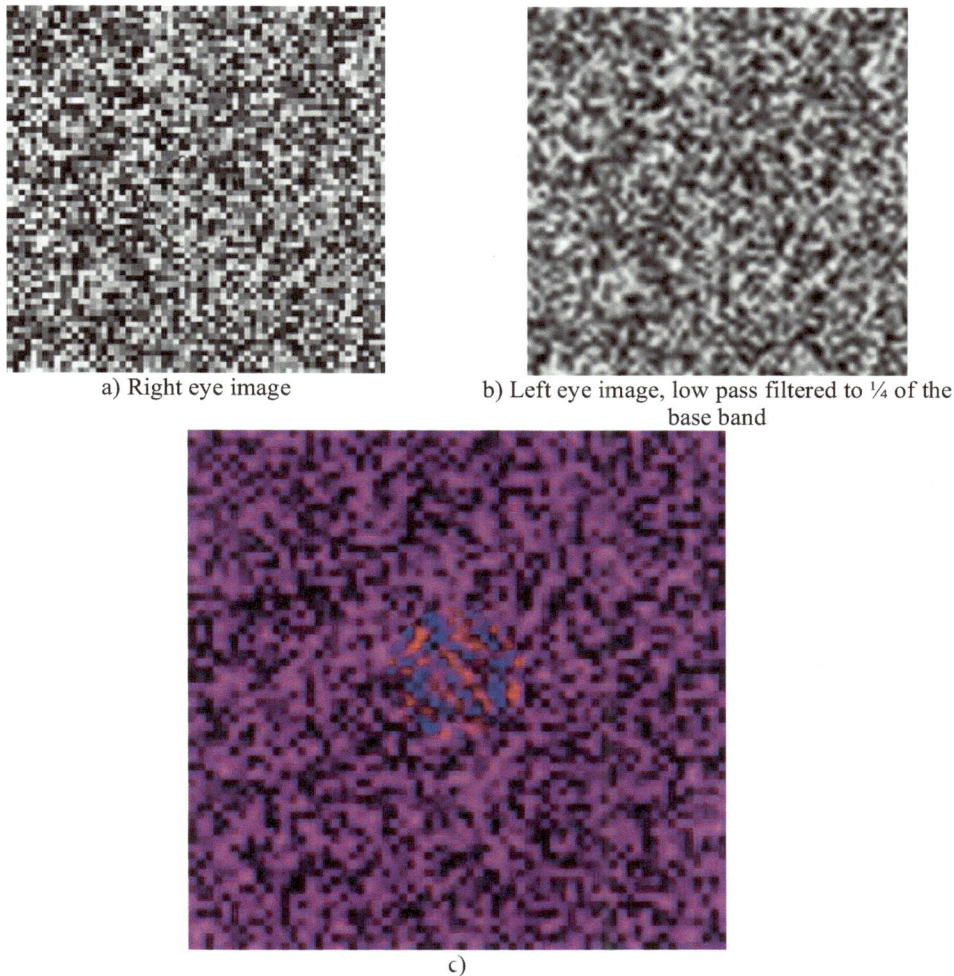

**Fig. 8.7.** Examples of computer generated random dot patterns used in the experiments: a), b) – right and left images for stereoscope or crossed eyes view, c) –anaglyph (left eye–red, right eye- blue). Right images exemplify low pass filtering to ¼ of the image base band in each dimension. When images are fused using stereoscope or crossed eyes and, for the anaglyph, color red-blue glasses, a test circle 3-D target can be seen without visible losses of the image resolution

In the experiments on the accuracy of parallax measurement, a test target with a moderate parallax was randomly placed within the image and viewers were requested to place the computer generated cross mark to the center of the target as soon as it is detected. Coordinates of the mark were compared by the computer program with actual coordinates of the target and the localization error was calculated. Each measurement was repeated, for the same image blur, several tens of times for statistical estimation of the localization error mean value and standard deviation that were used to characterize the parallax measurement accuracy. Note that parallax values in these experiments were set to values higher than the threshold values found in the previous experiments.

For image blur, low pass filtering was implemented in the domain of Discrete Fourier Transform. The filters were specified by the fraction, from 1 to 0.1 with a step of 0.1, of the image bandwidth. The degree of blur specified by filter bandwidth directly translates into the reduction of the number of samples required to generate the blurred image.

Typical results of the experiments are summarized in Figs. 8.8-8.9. Both detection threshold and standard deviation of localization error in these figures as well as depth values indicated in Figs. 8.9 and 8.10 are given in terms of the parallax values introduced for test object, the values being measured in units of inter-pixel distance, such that threshold or error equal to 1 correspond to the parallax in one sampling interval.

Figs. 8.8(a), 8.8(b), represent data obtained in experiments on stereopsis threshold for two test images, random dots and color random patches carried out with image low pass filtering. Blur factor in these graphs indicates fraction of the image base band left by the low pass filter: blur factor one corresponds to no low pass filtering, blur factor two corresponds to low pass filtering to half of the base band, three - low pass filtering to one third of the base band, and so on. For random dots test image, results of 10 experiments for one viewer and their mean and median are shown to illustrate typical data spread in experiments with one viewer. For random patches image, average data for 3 viewers are shown along with their mean and median to illustrate spread of averaged data for several viewers. Figs. 8.8 (c), 8.8 (d) represent stereopsis threshold data obtained for random dot test image using for image blur Haar and Daubechis-1 wavelet low pass filters, correspondingly. In these figures, "Scale index" specifies the scale in wavelet multi-resolution decomposition: scale index one corresponds to the initial resolution, scale index two corresponds to half of the initial resolution, scale index three corresponds to quarter of the initial resolution, four – to one eight of the initial resolution, and so on.

Figure 8.9 represents results of evaluating, using a stereoscope and two-color, 3D target localization accuracy obtained for different test images, using low pass filtering of one of two stereo images.

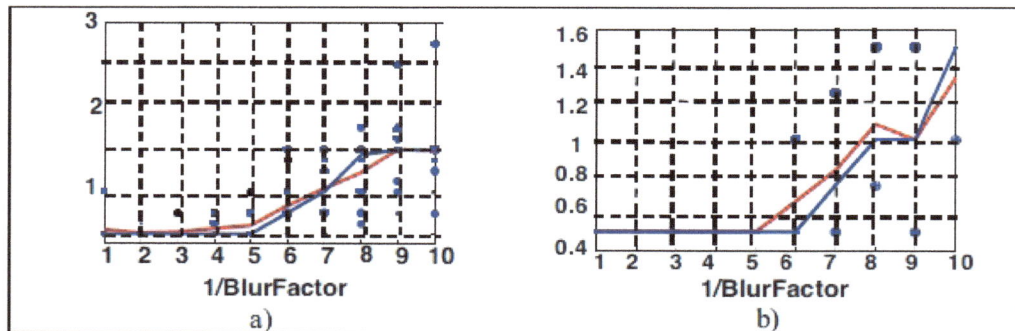

**Fig. 8.8.** Stereopsis threshold as a function of image blur: a) test image "random dots", case "ideal low pass filter, 10 experiments (dots) with one viewer and data average (red) and median (blue); b) - test image "Color random patches", case "low pass filter, 3 viewers (dots) and data average (red) and median (blue)

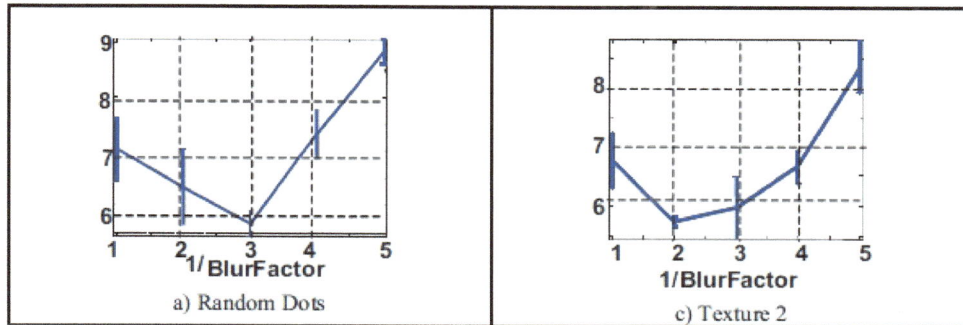

**Fig. 8.9.** Standard deviation of 3D target localization error as a function of image blur by means of low pass filtering for two test image; data spread is shown by vertical bars

The results of these experiments clearly show that both threshold and target localization accuracy do not substantially suffer from blur of one of two images up to the blur that corresponds to 5x5 to 7x7 times reduction in the number of pixels of one of two stereo images. For blur factor larger then 1/8-1/10, rapid growth of the stereopsis threshold and of the localization accuracy was observed.

Refs. [10] reports similar experiments with similar results demonstrating tolerance of human visual system to blur depth map. Experiments with quantization of depth maps of stereo images reported in Ref. [11] evidence also that coarse quantization of depth maps of stereo images to about 20-30 levels does not noticeably affects 3D perception.

## 8.4    Computer-generated display holograms

Several solutions that are computationally inexpensive and at the same time are quite sufficient for creating 3D visual sensation with synthetic display holograms have been suggested ([2]): multiple view compound macro-holograms, composite stereo-holograms and "programmed diffuser holograms".

### 8.4.1    *Multiple view compound macro-holograms.*

In this method, the scene to be viewed is described by means of multiple view images taken from different directions in the required view angle, and, for each image, a hologram is synthesized with an account of its position in the viewing angle (see Fig. 8.10). Each hologram has to be, approximately, of the size of the viewer's eye pupil. These elementary holograms will reconstruct different aspects of scenes from different directions, which are determined by their position in the view angle. The set of such holograms is then used to build a composite, or mosaic, macro-hologram.

It is essential that, for scenes given by their mathematical models, well-developed methods and software/hardware instrumentation tools of the modern computer graphics can be used for fast generating multiple view images needed for computing elementary holograms. As for the computational complexity of the synthesis of composite holograms, it can be estimated as following. If individual images are of $N \times N$ -pixels resolution, the complexity of the synthesis of elementary holograms with the use of fast Fourier Transforms is of the order of $N \log N$ operation.

Therefore, the computational complexity of synthesis of the composite hologram composed of $M \times M$ elementary holograms, is of the order $M^2 N \log N$ operations. Note that the computational complexity of generating a single hologram of the same size is $O\left( M^2 N \log M^2 N \right)$ which is $\log M^2 N / \log N$ times higher.

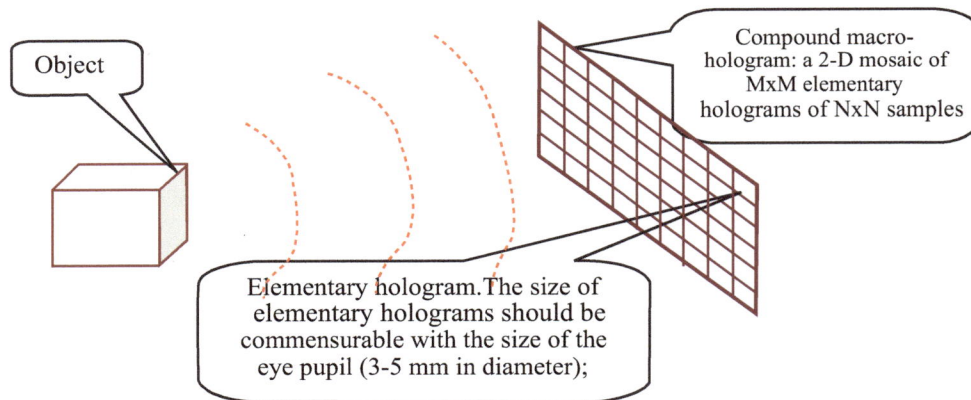

Fig. 8.10. Principle of the synthesis of composite holograms

In Refs. [10], an experiment on the synthesis of such a composite macro-hologram composed of 900 elementary holograms of 256x256 pixels was reported. The hologram contained 30x30 views, in spatial angle $-90^{\circ} \div 90^{\circ}$, of an object in a form of a cube. The synthesized holograms were encoded as kinoforms and recorded with sample size 12.5 mcm. The physical size of elementary holograms was 3.2x3.2 mm. Each elementary hologram was repeated, in the process of recording, 7x7 times to the size 22.4x22.4 mm. The size of the entire hologram was 672x672 mm2. Being properly illuminated, the hologram can be used for viewing the reconstructed scene from different angles, as, for instance, through a window (Fig. 8.11, left). Looking through the hologram with two eyes, viewers are able to see 3D image of a cube (Fig. 8.11, right) floating in the air.

Fig. 8.11. Viewing compound computer-generated hologram (left) and one of the views reconstructed from the hologram (right). The entire macro-hologram was composed of 900 elementary kinoform holograms of 256x256 pixels, which corresponded to 30x30 views in spatial angle $-90^{\circ}$ - $+90^{\circ}$. The holograms were recorded with pixel size 12.5 mcm. Size of the elementary hologram was 3.2x3.2 mm. Each elementary hologram was repeated 7x7 times to the size 22.4x22.4 mm.

### 8.4.2   Composite stereo-holograms.

A special case of multiple view mosaic macro-holograms is composite stereo-holograms. Composite stereo holograms are synthetic Fourier holograms that reproduce only horizontal parallax ([4, 11, 12]). When viewed with two eyes as through a window, the holograms are capable of creating 3D sensation thanks to stereoscopic vision. With such holograms arranged in a circular composite hologram, full 360 degrees view of the scene can be achieved. Fig. 8.12 shows such a hologram and examples of images from which it was synthesized ([4,11]). The entire hologram was composed of 1152 fragmentary kinoform holograms of 1024x1024 pixels recorded with pixel size 12.5 mcm. The total size of the hologram was 240 cm. In stationary state of the hologram, viewer looking through the hologram from different positions was able to see a 3D image of an object in a form of a "molecule" of six "atoms" differently arranged in space. When the hologram was rotated, the viewer was able to see "atoms" continuously rotating in space and easily recognize the rotation direction.

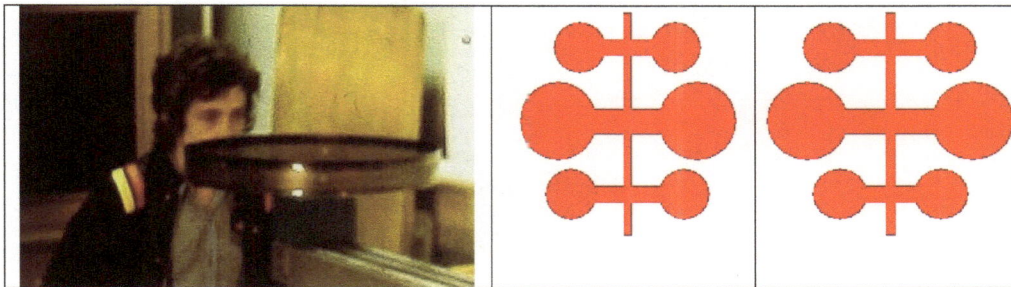

**Fig. 8.12.** Synthetic computer-generated circular stereo macro-hologram (left) and two views of the scene (right)

### 8.4.3   "Programmed diffuser" holograms.

The "Programmed diffuser" method for synthesis of Fourier display hologram was suggested for generating digital holograms capable of reconstructing different views of 3-D objects whose surfaces scatter light diffusely ([3, 10]). This method assumes that objects are specified in the object coordinate system $(x, y, z)$ by their "macro" shape $z(x, y)$, by the magnitude of the object reflectivity distribution $A(x, y)$ in the object plane $(x, y)$ and by the directivity pattern of the diffuse component of its surface. The diffuse light scattering from the object surface is simulated by assigning to the object a pseudo-random phase component (a "programmable diffuser"), whose correlation function corresponds to the given directivity pattern of the object surface. An algorithm for generating pseudo-random number sequences with a prescribed correlation function is given in Ch. 5, Sect. 5.5. This pseudo-random phase component is combined with the deterministic phase component defined by the object macroscopic shape to form the distribution of the phase of the object wavefront. A flow diagram of synthesis of "programmed diffuser" holograms is presented in Fig. 8.13.

Holograms synthesized with this method exhibit spatial inhomogeneity that is directly determined by the geometrical shape and diffuse properties of the object surface. This allows imitation of viewing the object from different direction by means of reconstruction of different fragments of its "programmed diffuser" hologram as it is illustrated in Fig. 8.14 on an example of a cone shaped object.

Fig. 8.15 presents frames of a movie that show reconstruction of a programmed diffuser hologram of a cone-shaped (left) and hemisphere-shaped uniformly painted object. One can clearly see a play of light patches on the objects in different viewing positions associated with different fragments of holograms. Fig. 8.16 presents one more simulation example of reconstruction of a programmed diffuser hologram of a hemisphere with 9 digits painted on its surface, and Fig. 8.17 illustrate optical reconstruction of computer generated programmed diffuser hologram of a pyramid and a hemisphere. Note in conclusion that for recording such holograms all encoding methods described in Ch. 6 can be used including kinoform.

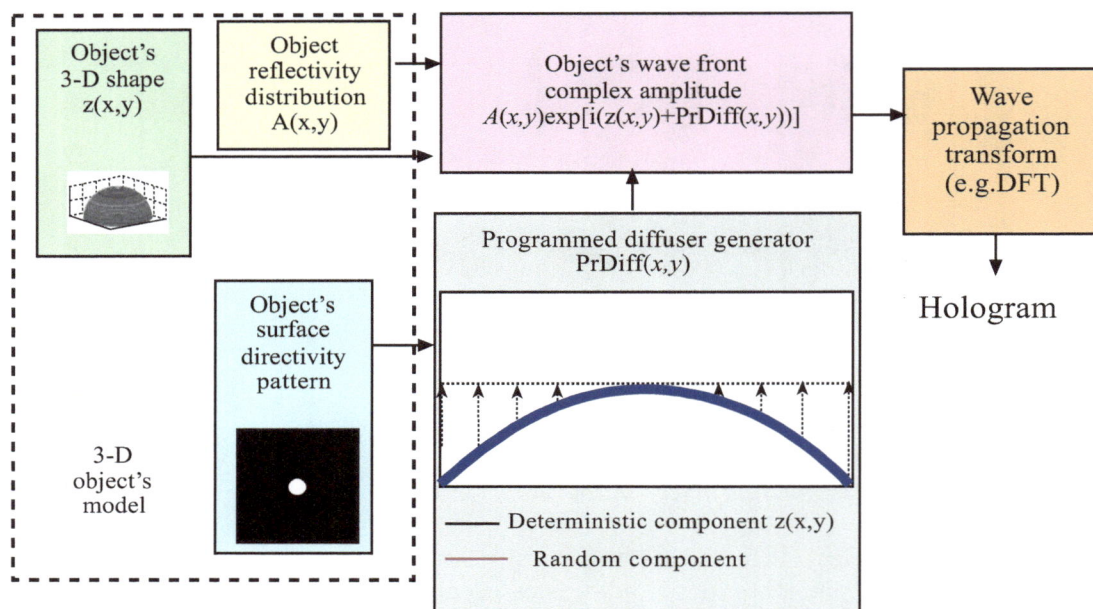

**Fig. 8.13**. Block-diagram of synthesis of programmed diffuser holograms

## 8.5   Digital holographic displays

One of the major obstacles, if not the major one, for implementing 3D digital holographic visual communication is the lack of digital holographic displays, devices for converting data on the distribution of the amplitude and phase in the hologram plane into optical holograms for visual reconstruction of 3D images. One can distinguish two types the holographic displays: static displays for reproducing holograms of static scenes similarly to printers in conventional imaging and dynamic ones for reproducing holograms of dynamic scenes. While creation of dynamic digital holographic displays still remains quite problematic, static digital holographic displays are becoming feasible.

Important practical issues in the design of devices for displaying computer generated display holograms are sources of light for reconstruction of the holograms and encoding of holograms for displaying color 3D images. In what follows we discuss some possible solutions.

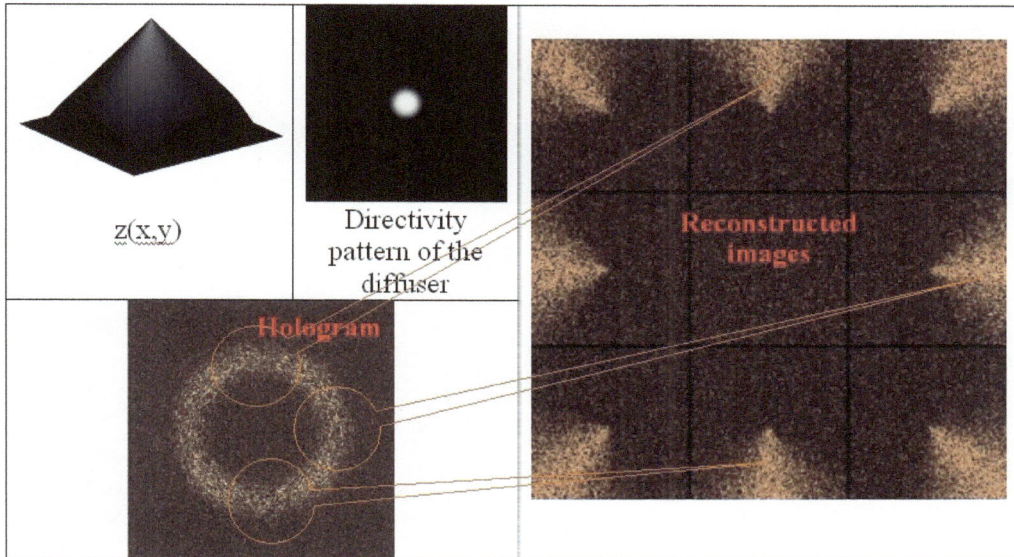

**Fig. 8.14.** Programmed diffuser hologram of a cone-shaped object and results of its reconstruction from its different fragments

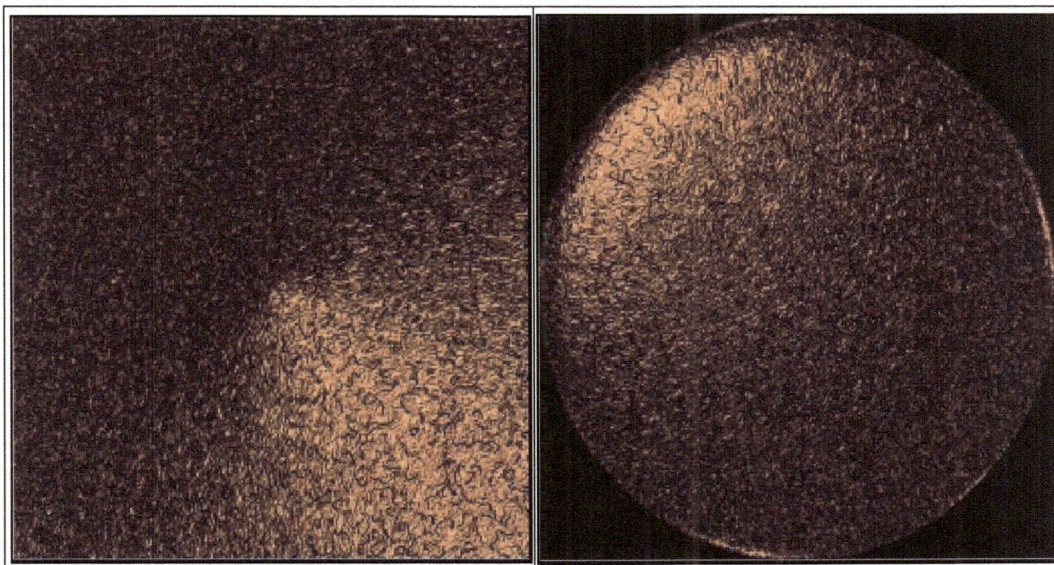

**Fig. 8.15.** Frames of a movie that demonstrate computer simulated reconstruction of programmed diffuser holograms of cone-shaped (left) and hemisphere-shaped objects (right, click to animate).

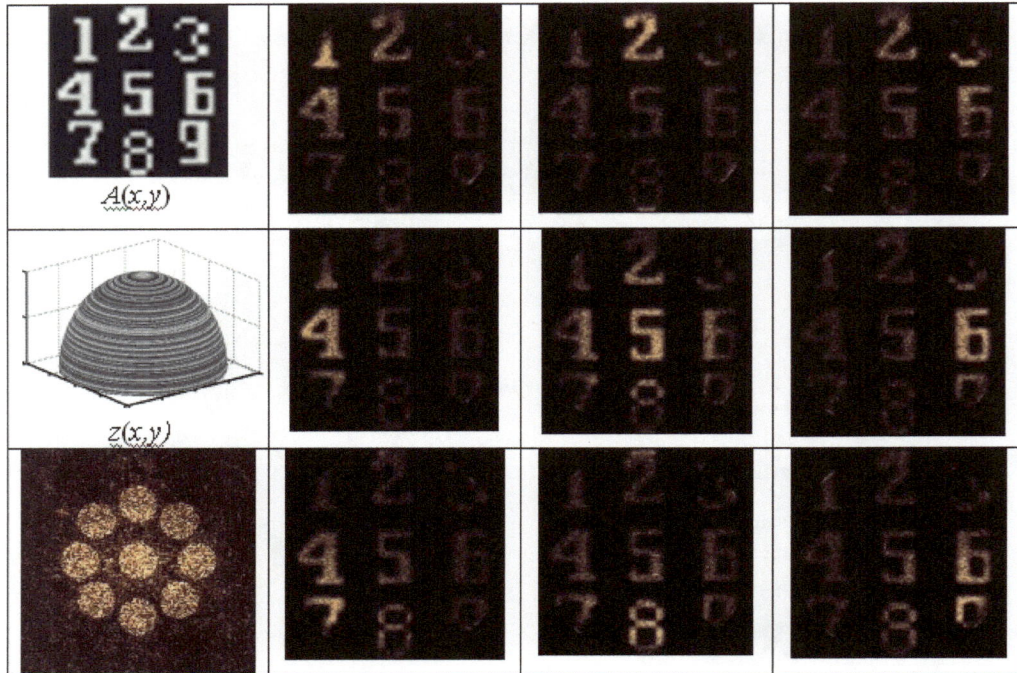

**Fig. 8.16.** Object's image (upper left), object's shape (center left), its programmed diffuser hologram (bottom left; circles on the hologram highlight different reconstruction windows) and nine images reconstructed from northwest, north, northeast, west, center, east, southwest south and southeast fragments of the hologram (right).

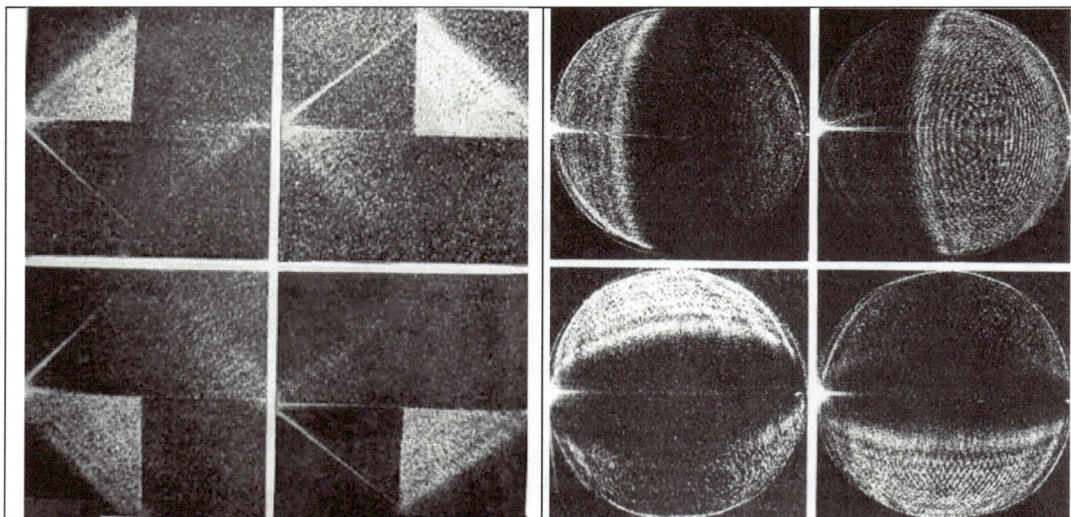

**Fig. 8.17.** Examples of optical reconstruction of programmed diffuser holograms of objects in form of a pyramid (left) and hemisphere (right).

### 8.5.1 Low coherence light and white light computer-generated holograms. Hybrid optical-digital holograms.

As computer-generated holograms reproduce distribution of the object wave front amplitude and phase, computed for a certain light wavelength, it is a common belief that, for displaying computer-generated holograms, one needs to use a laser. However, high coherency lasers are not really absolutely necessary. The incoherence of the light beam used for reconstruction of holograms influences the accuracy in reconstruction of the phase of the hologram and may cause image blur due to different scaling of reconstructed images for different wavelengths of the light. In order to keep the image blur due to the imperfect coherence of the light on an acceptable low level, it is sufficient to use a reconstruction source of light with the degree of coherence, measured as the ratio of the light frequency to its spectrum spread, of the order of the largest, in horizontal or vertical dimension, number of pixels in the scene images (say, 1000). This is by 3-4 orders of magnitude lower than that needed for recording optical holograms. Obviously, for such incoherence, the accuracy in hologram phase reproduction will be of the order of 0.1%, which is also quite sufficient.

The use, for reconstruction of computer-generated holograms of such sources of low coherence quasi-monochromatic light can simplify the design of the digital holographic displays and, in addition, is beneficial in terms of reducing reconstruction speckle noise.

Ideally, the use, for reconstruction of hologram, of natural white light would be the most appropriate. The great advantage of Denisyuk-type optical holograms is that that they do not need, for reconstruction, a coherent light of any coherence and are reconstructed in white light. This property will also be very beneficial for computer-generated holograms, especially for static holographic displays. At least three methods can be suggested for producing computer-generated holograms capable of reconstruction in white light:

1. Recording computer-generated hologram onto optical media pre-exposed by a Denisyuk-type hologram of the reconstruction source of light.
2. Making "sandwich-holograms" out of computer-generated holograms of the object to be displayed and of a Denisyuk-type hologram of the reconstruction source of light.
3. Holographic copying onto a Denisyuk-type hologram of an image reconstructed, in coherent light, from the computer-generated hologram

When such holograms are illuminated by white light, their Denisyuk type component plays a role of the needed light source.

### 8.5.2 Computer generated colour display holograms

Of course, computer-generated display holograms must ultimately be capable of reproducing color images. Color CG-holograms can be generated as color-separated RGB holograms and recorded by interlacing their corresponding samples covered by the respective RGB-filters. Fig. 8.18 illustrates one possible hologram encoding scheme for recording color computer generated holograms on a phase media. The scheme is an extension of the double-phase encoding method ([3, 4, 14], see Ch. 5,

Sect. 5.2). The use of phase media for recording computer-generated holograms is advantageous is terms of hologram light efficiency.

For encoding the amplitude and phase of RGB samples of the computed object wavefront in the hologram plane, groups of three cells arranged in the hexagonal sampling grid are used per each RGB component, each group for one of R, G and B sample (see Fig. 8.18, a).

Each of RGB cell groups is covered by corresponding *R*, *G* or *B* color filters. Each *k*-th cell in the *n*-th *R*-, *G*- or *B*- group modulates phase of the illuminating beam by introducing phase shift $\varphi_{R,n}^{(k)}$, $\varphi_{G,n}^{(k)}$ and $\varphi_{B,n}^{(k)}$, correspondingly. The phase shifts $\varphi_{R,n}^{(k)}$, $\varphi_{G,n}^{(k)}$ and $\varphi_{B,n}^{(k)}$ for encoding of *n*-th sample of the corresponding hologram color component with amplitude $A_{C,n}$ and phase $\Phi_{C,n}$ are found, according to the drawing in Fig. 8.18, b), from the relationships:

$$A_{C,n}\exp\left(i\Phi_{C,n}\right) = A_0\left[\cos\left(\Phi_{C,n}-\varphi_{C,n}^{(1)}\right)+1+\cos\left(\varphi_{C,n}^{(3)}-\Phi_{C,n}\right)\right]$$

$$\varphi_{C,n}^{(3)} = -\varphi_{C,n}^{(1)},$$

where $C = R,G,B$.

Note, that such color holograms can be viewed in white light provided RGB- filters are appropriately narrow-band.

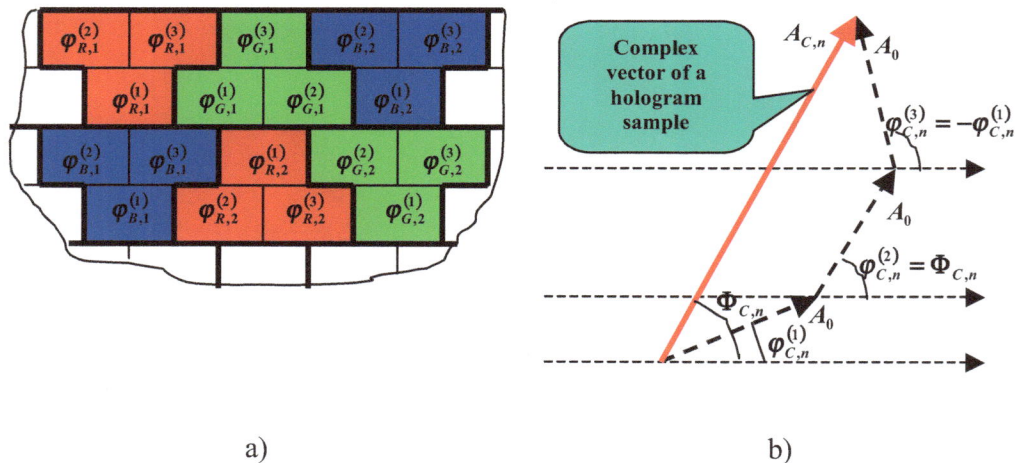

a)                                                                b)

**Fig. 8.18.** A hexagonal arrangement of the hologram samples on a phase media (a) and the three-phase method of hologram encoding for recording on phase media (b)

## References

1.    E.N. Leith, J. Upatnieks, New techniques in Wavefront Reconstruction, JOSA, v. 51, 1469-1473, 1961
2.    Yu. N. Denisyuk, Photographic reconstruction of the optical properties of an object in its own scattered radiation field, Dokl. Akad. Nauk SSSR, 144,1275-1279, 1962
3.    S.H. Lee, in: Progress in Optics, Ed. E. Wolf, 1978, v. 16, p. 119-132

4.    L. Yaroslavsky, N. Merzlyakov, Methods of Digital Holography, Consultance Bureau, N.Y., 1980

5.    B. Julesz, Foundations of Cyclopean Perception, (The University of Chicago Press, 1971), p. 96,

6.    L. P. Yaroslavsky, "On Redundancy of Stereoscopic Pictures," Acta Polytechnica Scandinavica, n. 149. Image Science'85. Proceedings. Helsinki, Finland., 11-14 June 1985, V. 1, p. 82-85

7.    L. Yaroslavsky, Digital Holography and Digital Image Processing, Kluwer Academic Publishers, Boston, 2004

8.    L.P. Yaroslavsky, J. Campos, M. Espinola, I. Ideses, Optics Express, v. 13, No. 26, Dec. 22, 2005, p. 10895

9.    I. Ideses, L. P. Yaroslavsky. B. Fishbain, "How Sharp Must Depth Maps be For Good 3D Video Synthesis: Experimental Evaluation and Applications", Digital Holography and Three-Dimensional Imaging 2007 Conference, Vancouver, BC, Canada, June 18-20, 2007

10.   I. Ideses, L. Yaroslavsky, I. Amit, B. Fishbain, "Depth Map Quantization – How Much Is Sufficient?", 3DTV-Conference the true vision, capture, transmission and display of 3D video, 7-9 May, 2007, Kicc Conference Center, Kos Island, Greece.

11.   V.N. Karnaukhov, N.S. Merzlyakov, M.G. Mozerov, L.P. Yaroslavsky, L.I Dimitrov,E. Wenger, Digital Display Macro-holograms. In: Computer Optics, ISSN 0134-2452, Ed. Ye. P. Velikhov and A.M. Prokhorov, Issue 14-15, part 2, Moscow, 1995

12.   V.N. Karnaukhov, N.S. Merzlyakov, L.P. Yaroslavsky, Three dimensional computer generated holographic movie, Sov.Tech. Phys. Lett., v.2, iss.4, Febr. 26, 1976, p.169-172.

13.   Yatagai T. Stereoscopic approach to 3-D display using computer-generated holograms, Appl. Opt., 1976, v. 15, No 11, p. 2722-2729

14.   N.S. Merzlyakov, L.P. Yaroslavsky, Simulation Of Light Spots on Diffuse Surfaces by Means of a "Programmed Diffuser", Sov. Phys. Tech. Phys., v. 47, iss.6, 1977, p. 1263-1269.

15.   C.K. Hsuen, A.A., Sawchuk, Computer generated double phase holograms, Appl. Optics., 1978, v. 17, No 24, p.3874-3883

## 9      Image Processing Methods

**Abstract:** Work in digital holography always involves image processing, such as image denoising, deblurring, resampling, geometrical transformation and alike. In this chapter, basic approaches and methods of image processing most relevant to applications of digital holography are reviewed.

### 9.1      Stochastic Noise Models in Imaging and Digital Holography

#### *9.1.1      Mathematical models of imaging and holographic systems*

Imaging systems always have certain technical limitations in their design and implementations and generate images that are not as perfect as they would be if there were no implementation limitations. Imperfections of real images may be treated as caused by distortions introduced by imaging systems to hypothetical perfect, or "ideal" signals. Correction of these distortions is the primary goal of image processing.

Methods for distortion correction are based on the canonical model of imaging systems shown in Fig. 9.1.

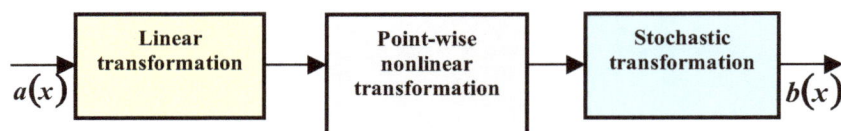

**Fig. 9.1.** A canonical model of imaging systems

The model represents image formation as a combination of signal linear transformations, point-wise nonlinear transformations and stochastic transformations that are applied to a hypothetical perfect image $a(x)$ and jointly determine the system's output image $b(x)$.

***Linear transformations*** are specified in terms of the imaging system ***point spread function*** or frequency response, or its Fourier transform called "***frequency response***". Frequency response of the ideal imaging system is assumed to be uniform for all frequencies in a certain base band. Frequency responses of real imaging usually more or less rapidly decay at high frequencies, which results in image distortions such as image blur.

Point-wise nonlinear are specified in terms of the system transfer functions. It is assumed that the ideal system has a linear transfer function. Deviations of system transfer functions from the linear ones cause image ***gray scale nonlinear distortions***.

***Stochastic transformations*** model signal random distortions that cause random interferences, or noise, in output images. Stochastic transformations are specified by statistical noise models described in Sect. 9. 2.

The processing goal is estimating the perfect signal $a(x)$ given distorted signal $b(x)$ produced by the system. This problem is frequently referred to as the ***inverse problem***. Signal processing applied to system's output signals $\{b(x)\}$ to produce an estimate $\hat{a}(x)$ of the "ideal" signal is called ***signal recovery*** or, in application to image processing, ***image restoration***. Similar problem is ***image reconstruction***. This term usually refers to such imaging methods as tomography and holography.

If the system does not introduce any random distortions and system's parameters such as point spread function and transfer function are known, the inverse problem has a trivial solution: signal $b(x)$ has to be subjected to transformations inverse to those introduced by the system. However, one cannot, in general, neglect random or other uncontrolled distortions that always happen. In reality, applying inverse transformation may result in artifacts that may even make total job useless. For solving the inverse problem in such cases, the statistical approach that explicitly accounts for signal random distortions appears to be the most appropriate.

Generally, signal recovery, image restoration and reconstruction is an indivisible procedure that should account for and correct all distortions. However, in practice this process is divided into separate steps carried out in the order inverse to that distortion factors have in the model.

### 9.1.2   Statistical models of stochastic transformations

The most useful models to describe random interferences in imaging systems are those of additive, multiplicative, Poisson noise, impulse and speckle noise models.

#### 9.1.2.1   Additive noise model

The ***additive signal independent noise model*** (ASIN-model) assumes signal transformation described as

$$b(x) = a(x) + n(x), \tag{9.1.1}$$

where $a(x)$ is a perfect signal, $b(x)$ is an imaging system output signal and $n(x)$ is a random process statistically independent on signal $a(x)$ and referred to as ***additive noise***. The assumption of statistical independence assumes that statistical averaging $AV\{\cdot\}$ applied to signal $b(x)$ and to results of its transformations can be carried out on statistical ensembles of signals $\Omega_S$ and noise $\Omega_N$ separately:

$$AV_{\{\Omega_S, \Omega_N\}}\{b(x)\} = AV_{\Omega_S}\{a(x)\} + AV_{\Omega_N}\{n(x)\}. \tag{9.1.2}$$

For ASIN-model, mean value $\bar{n} = AV_{\Omega_N}\{r(x)\}$ of noise is of no concern and it is natural to assume that $\bar{n} = 0$. With this assumption, mean value of the signal corrupted by additive zero mean noise is equal to the uncorrupted signal and standard deviation of its fluctuations is equal to the standard deviation of noise:

$$AV_{\Omega_N}\{b(x)\} = a(x); . \tag{9.1.3}$$

$$\sqrt{AV_{\Omega_N}\{|b(x) - a(x)|^2\}} = \sqrt{AV_{\Omega_N}\{|n(x)|^2\}}. \tag{9.1.4}$$

Most common statistical model for additive noise is that of Gaussian random process. Gaussian random process is completely statistically determined by its probability density called the **normal distribution**

$$p(n) = \frac{1}{\sqrt{2\pi}\sigma_n} \exp\left[-\frac{(n-\bar{n})^2}{2\sigma_n^2}\right], \tag{9.1.5}$$

where $\sigma_n$ is its standard deviation and $\bar{n}$ is its mean value, which, for ASIN-model, is usually assumed to be zero, and by its **correlation function**

$$R_n(x_1, x_2) = AV_{\Omega_N}\{n(x_1) \cdot n(x_2)\}. \tag{9.1.6}$$

According to the **central limit theorem** of the probability theory, normal distribution is a good model for probability density distribution of random values that are obtained as a sum of very many statistically independent random variables. For instance, it is a good model of the distribution of **thermal noise** caused by random fluctuations of velocities of electrons in electron current in signal sensors.

Random process is a **stationary process**, if its correlation function depends only on the distance between points $x_1$ and $x_2$: $R_n(x_1, x_2) = R_n(x_1 - x_2)$. Fourier transform

$$N(f) = \int_{-\infty}^{\infty} R_n(x) \exp(i2\pi fx) dx. \tag{9.1.7}$$

of correlation function of a stationary random process is called its **spectral density**, or **power spectrum**. Alternatively, spectral density of a stationary random process can be defined as average, over all realizations of the process $\{n(x)\}$, of squared module of Fourier spectra of the realizations

$$N(f) = AV_{\Omega_N}\left\{\left|\int_{-\infty}^{\infty} n(x) \exp(i2\pi fx) dx\right|^2\right\}. \tag{9.1.8}$$

Random process is called **uncorrelated** if its spectral density is a constant, or respectively, its correlation function is a delta-function:

$$N(f) = N_0 = \sigma_n^2;$$
$$R_n(x) = N_0 \delta(x) = \sigma_n^2 \delta(x). \quad\quad\quad (9.1.9)$$

Uncorrelated Gaussian random process is conventionally called **white noise**.

In some of applications, additive noise interferences are highly spatially correlated. It is most convenient to describe the presence of correlations in random processes of noise as concentration of noise power spectrum in certain areas of frequency domain. If noise power spectrum is concentrated only around a few isolated points in frequency domain, this type of additive noise is called **moiré noise**.

### 9.1.2.2 Multiplicative noise model

**Multiplicative noise** (MN-) model assumes signal transformation defined as

$$b(x) = m(x) \cdot a(x), \quad\quad\quad (9.1.10)$$

where $m(x)$ is signal independent random process with mean value equal to 1. Similarly to mean value of a signal corrupted by an additive zero mean noise, mean value of a signal corrupted by multiplicative noise is also equal to the mean value of the non-corrupted signal:

$$AV\{b(x)\} = AV_{\Omega_S}\{a(x)\} \cdot AV_{\Omega_N}\{n(x)\} = AV_{\Omega_S}\{a(x)\}. \quad\quad\quad (9.1.11)$$

The second moment of the corrupted signal is the product of the second moment of non-corrupted signal and the variance of noise:

$$AV\{b^2(x)\} = AV_{\Omega_S}\{a^2(x)\} \cdot AV_{\Omega_N}\{n^2(x)\}. \quad\quad\quad (9.1.12)$$

For multiplicative noise, the ratio of the corrupted signal variance $AV\{b^2(x)\}$ to the variance of the uncorrupted signal $AV_{\Omega_S}\{a^2(x)\}$ is equal to the variance of the multiplicative noise $AV_{\Omega_N}\{n^2(x)\}$ and does not depend on the signal mean value. This is a characteristic property of the multiplicative noise.

### 9.1.2.3 Poisson noise model

**Poisson noise model** is used for describing quantum fluctuations in sensors such as fluctuations of number of photons or electrons in CCD or CMOS photo and X-ray sensitive sensors. For the Poisson noise model, signal values are nonnegative integer numbers that have **Poisson distribution** probability density:

$$P(q) = \frac{\overline{q}^q}{q!} \exp(-\overline{q}), \quad\quad\quad (9.1.13)$$

where $\overline{q} = AV(q)$ is the mean value of $q$. It is the property of Poisson distribution that variance $Var(q) = AV(q^2) - AV(\overline{q}^2)$ of $q$ is also equal to $\overline{q}$:

$$\left( AV\left\{ \left( q - \overline{q} \right)^2 \right\} \right)^{1/2} = \overline{q} \; .$$

(9.1.14)

For $\overline{q} \rightarrow \infty$, Poisson distribution, according to Moivre-Laplace theorem, tends to normal distribution with standard deviation $\sqrt{\overline{q}}$

$$P(q) = \frac{1}{\sqrt{2\pi\overline{q}}} \exp\left[ -\frac{\left( q - \overline{q} \right)^2}{2\overline{q}} \right].$$

(9.1.15)

For Poisson distributions, ratio of standard deviation to mean value tends to zero with growth of the mean value:

$$\lim_{\overline{q}\rightarrow\infty}\left\{ \frac{\left[ AV_q \left| q - AV_q \left( q \right) \right|^2 \right]^{1/2}}{AV_q \left( q \right)} \right\} = \lim_{\overline{q}\rightarrow\infty}\left\{ \frac{\sqrt{\overline{q}}}{\overline{q}} \right\} = 0 \; ,$$

(9.1.16)

which means that with growth of $\overline{q}$ fluctuations are becoming less and less noticeable.

### 9.1.2.4   Impulse noise model

***Impulse noise model*** assumes that signal random interferences are caused by two independent random factors: by a binary random process $e(x)$ that assumes two values, zero and one, and by an additive noise $n(x)$:

$$b(x) = \left[ 1 - e(x) \right] \cdot a(x) + e(x) n(x)$$

(9.1.17)

Impulse noise model is applicable to discrete signals. It is characteristic for discrete signal transmission systems with random faults. The process $e(x)$ is specified by the probability that $e(x) = 1$ (the probability of error, or the probability of signal loss). Statistical properties of noise $n(x)$ can be arbitrary. If random process $n(x)$ takes only two values equal to minimum and maximum of $a(x)$, it is, in image processing jargon, referred to as ***pepper and salt noise***.

Obviously, visibility of noise and its interfering capability grows with its variance. However the noise variance is not the only statistical characteristic of noise that defines its interfering capability. Other important characteristics are the type of noise and its correlation properties. This is illustrated by examples of images distorted with additive and impulse noise shown in Fig. 9.2. The noise level in images is characterized in these images by the ratio PSNR of image dynamic range to standard deviation of noise. For impulse noise, standard deviation of noise is computed as standard deviation of the difference between initial and noisy images.

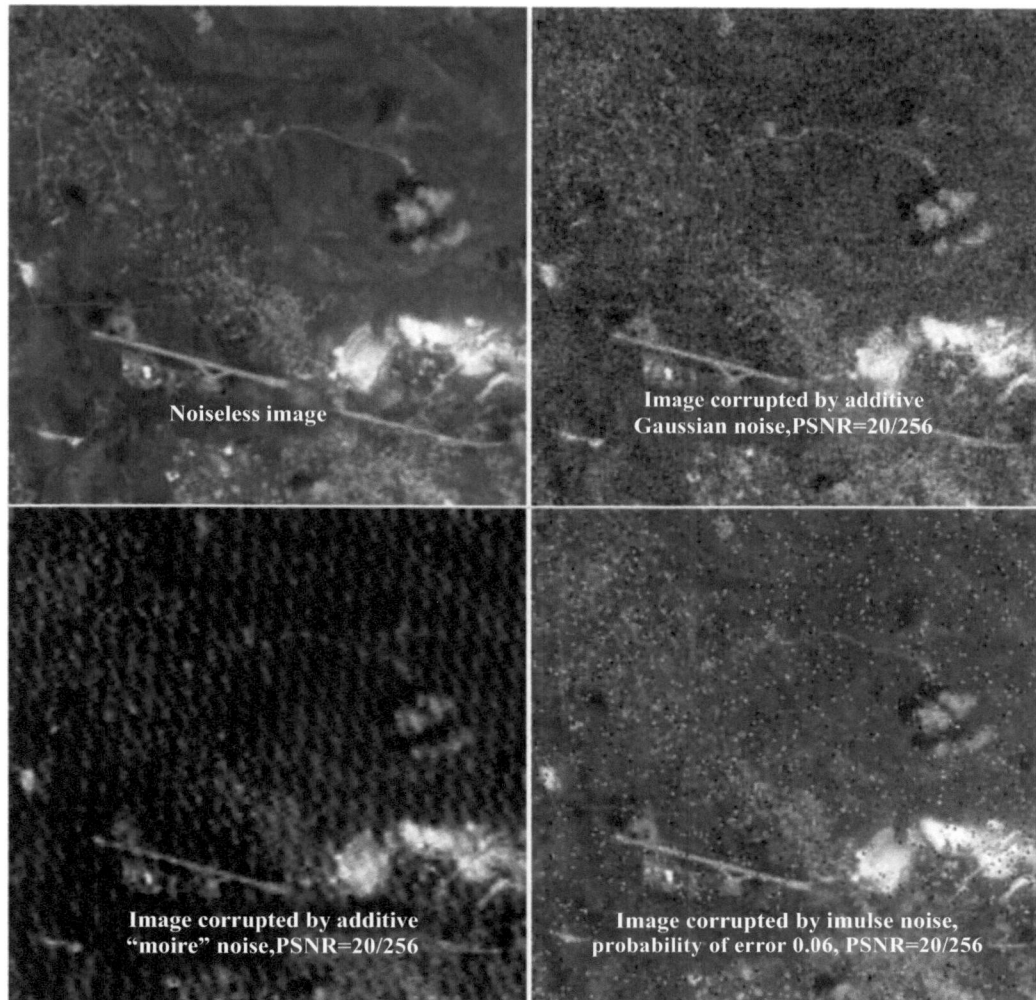

**Fig. 9.2.** Examples of images with additive and impulse noise. Observe different visibility of different types of noise with the same standard deviation

### 9.1.2.5    Speckle noise model

***Speckle noise*** model is used to describe chaotic patterns that are observed in images created by coherent imaging systems such as holographic, synthetic aperture radar and ultrasound imaging systems. This noise arises due to incapability of the imaging system to accurately reproduce microscopic variations of the object wavefront caused by properties of object surfaces to scatter radiation diffusely. Fig. 9.3 shows some examples.

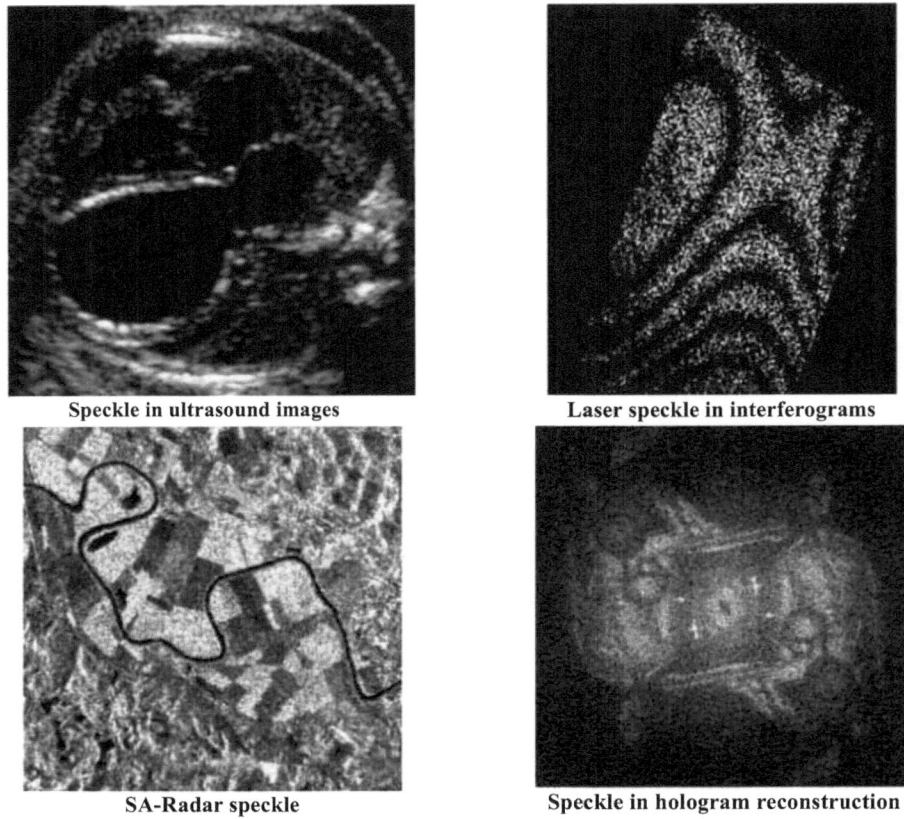

**Speckle in ultrasound images**

**Laser speckle in interferograms**

**SA-Radar speckle**

**Speckle in hologram reconstruction**

**Fig. 9.3.** Examples of speckle images

Most frequently, speckle noise is caused by insufficient resolving power of sensors, or, in holographic systems, by limitation of the area in which wavefront is measured by sensors or recorded on holograms. A model of speckle formation for this kind of distortions is shown in Fig. 9.4.

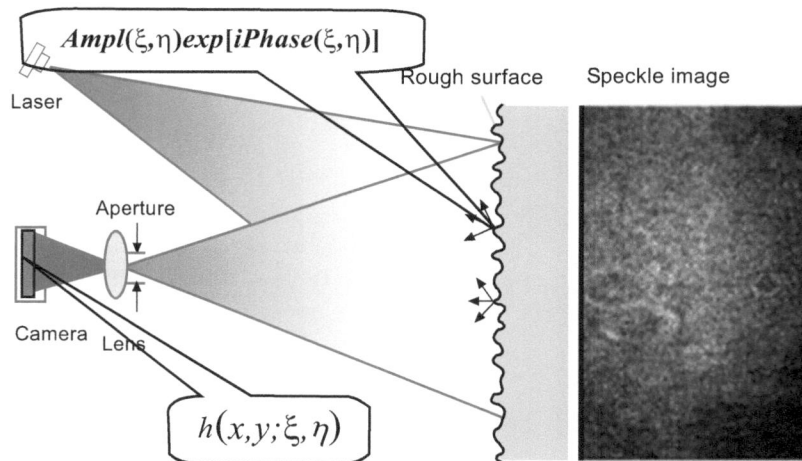

**Fig. 9.4.** A model of speckle formation

Let $a(\xi,\eta)\exp\left[i\theta(\xi,\eta)\right]$ be complex amplitude of the wavefront in the object plane that describes object reflectance/transmittance properties. For describing property of objects to diffusely scatter radiation, $\theta(\xi,\eta)$ can be modeled as a random process. Diffuse scattering uniformly in all direction corresponds to spatially uncorrelated process $\theta(\xi,\eta)$. Specular scattering takes place if $\theta(\xi,\eta)$ is a highly spatially correlated process. Fig. 9.5 illustrates examples of rough surfaces with different spatial directivity of light scattering and, correspondingly, different correlation functions, the Fourier transform of the spatial directivity function.

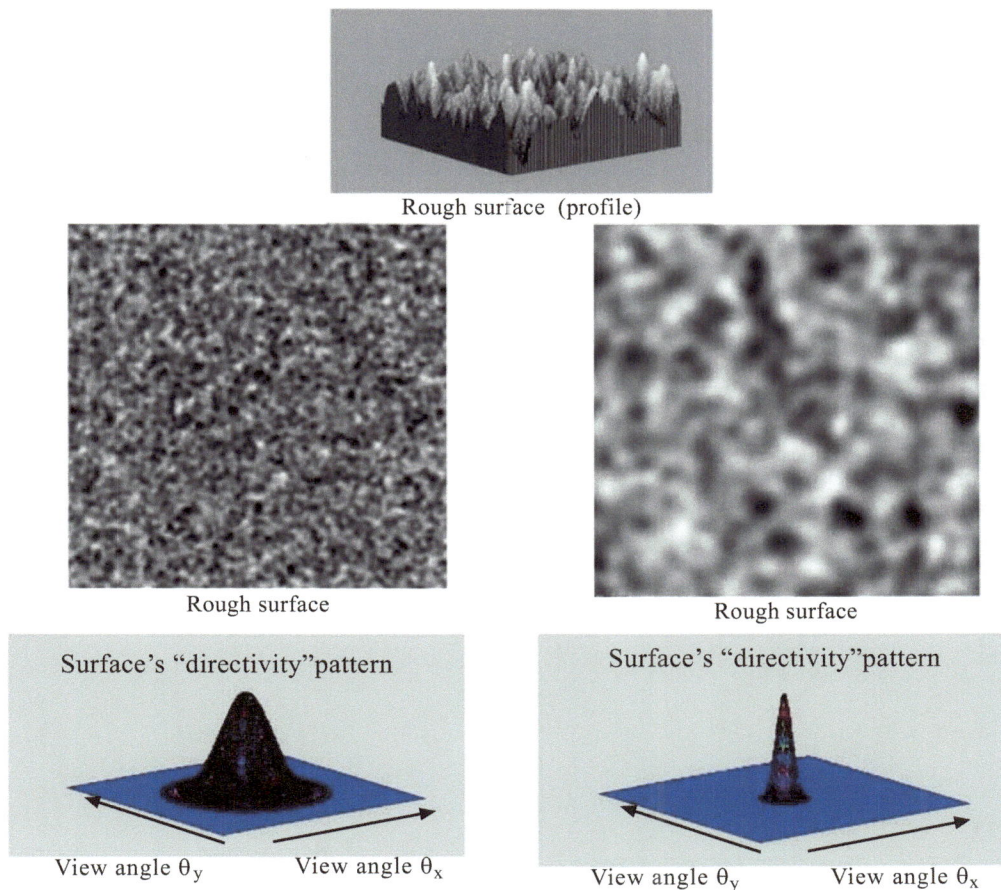

Rough surface  (profile)

Rough surface

Rough surface

Surface's "directivity"pattern

View angle $\theta_y$          View angle $\theta_x$

Surface's "directivity"pattern

View angle $\theta_y$          View angle $\theta_x$

**Fig. 9.5.** Examples of rough surfaces with different spatial directivity of scattering

Consider now a coherent imaging system with a point spread function $h(x,y;\xi,\eta)$. Then, for input object $a(\xi,\eta)\exp(i\theta(\xi,\eta))$, output image formed by a sensor sensitive to squared magnitude of the output wave front is

$$I(x,y)=\left|\int_{-\infty}^{\infty}\int_{-\infty}^{\infty} a(\xi,\eta)\exp\left[i\theta(\xi,\eta)\right]h(x,y;\xi,\eta)\,d\xi\,d\eta\right|^2. \tag{9.1.18}$$

Let the object has a uniformly painted surface such that $a(\xi,\eta) = A_0$. Then:

$$I(x,y) = \left| A_0 \int\limits_{-\infty}^{\infty} \int\limits_{-\infty}^{\infty} \exp\left[i\theta(\xi,\eta)\right] h(x,y;\xi,\eta) d\xi\, d\eta \right|^2 = \left| b^{re} \right|^2 + \left| b^{im} \right|^2 \qquad (9.1.19)$$

where

$$b^{re} = A_0 \int\limits_{-\infty}^{\infty} \int\limits_{-\infty}^{\infty} \cos\left[\theta(\xi,\eta)\right] h(x,y;\xi,\eta) d\xi\, d\eta, \qquad (9.1.20)$$

$$b^{im} = A_0 \int\limits_{-\infty}^{\infty} \int\limits_{-\infty}^{\infty} \sin\left[\theta(\xi,\eta)\right] h(x,y;\xi,\eta) d\xi\, d\eta \qquad (9.1.21)$$

As far as $\theta(\xi,\eta)$ is regarded as a random process, $b^{re}$ and $b^{im}$ are also random processes in coordinates $(\xi,\eta)$. Mean values of these processes is zero:

$$\overline{b^{re}} = AV_\theta\left(b^{re}\right) = A_0 \int\limits_{-\infty}^{\infty} \int\limits_{-\infty}^{\infty} AV_\theta\left\{\cos\left[\theta(\xi,\eta)\right]\right\} h(x,y;\xi,\eta) d\xi\, d\eta = 0 ;$$

$$\qquad (9.1.22,\text{a})$$

$$\overline{b^{im}} = AV_\theta\left(b^{im}\right) = A_0 \int\limits_{-\infty}^{\infty} \int\limits_{-\infty}^{\infty} AV_\theta\left\{\sin\left[\theta(\xi,\eta)\right]\right\} h(x,y;\xi,\eta) d\xi\, d\eta = 0 .$$

$$\qquad (9.1.22,\text{ b})$$

and their correlation functions are:

$$CF_{b^{re}}\left(\mathbf{x}_1,\mathbf{x}_2\right) = AV_\theta\left\{ b^{re}\left(\vec{\mathbf{x}}_1\right)\left[ b^{re}\left(\vec{\mathbf{x}}_2\right)\right]^* \right\} =$$

$$A_0^2 \int\limits_{-\infty}^{\infty} \int\limits_{-\infty}^{\infty} AV_\theta\left\{\cos\left[\theta(\vec{i}_1)\right]\cos\left[\theta(\vec{i}_2)\right]\right\} h(\vec{\mathbf{x}}_1;\vec{i}_1) h^*(\vec{\mathbf{x}}_2;\vec{i}_2) d\vec{i}_1\, d\vec{i}_2 =$$

$$A_0^2 \int\limits_{-\infty}^{\infty} \int\limits_{-\infty}^{\infty} CF_{c\theta}\left(\vec{i}_1,\vec{i}_2\right) h(\vec{\mathbf{x}}_1;\vec{i}_1) h^*(\vec{\mathbf{x}}_2;\vec{i}_2) d\vec{i}_1\, d\vec{i}_2 , \qquad (9.1.23,\text{ a})$$

$$CF_{b^{im}}\left(\vec{\mathbf{x}}_1,\vec{\mathbf{x}}_2\right) = AV_\theta\left\{ b^{im}\left(\vec{\mathbf{x}}_1\right)\left[ b^{im}\left(\vec{\mathbf{x}}_2\right)\right]^* \right\} =$$

$$A_0^2 \int\limits_{-\infty}^{\infty} \int\limits_{-\infty}^{\infty} AV_\theta\left\{\sin\left[\theta(\vec{i}_1)\right]\sin\left[\theta(\vec{i}_2)\right]\right\} h(\vec{\mathbf{x}}_1;\vec{i}_1) h^*(\vec{\mathbf{x}}_2;\vec{i}_2) d\vec{i}_1\, d\vec{i}_2 =$$

$$A_0^2 \int\limits_{-\infty}^{\infty} \int\limits_{-\infty}^{\infty} CF_{s\theta}\left(\vec{i}_1,\vec{i}_2\right) h(\vec{\mathbf{x}}_1;\vec{i}_1) h^*(\vec{\mathbf{x}}_2;\vec{i}_2) d\vec{i}_1\, d\vec{i}_2 , \qquad (9.1.23,\text{ b})$$

where $\vec{\mathbf{x}} = (x,y)$, $\vec{i} = (\xi,\eta)$ and $CF_{c\theta}\left(\vec{i}_1,\vec{i}_2\right)$ and $CF_{s\theta}\left(\vec{i}_1,\vec{i}_2\right)$ are correlation functions of $\cos\left[\theta(\xi,\eta)\right]$ and $\sin\left[\theta(\xi,\eta)\right]$. If functions $\cos\left[\theta(\xi,\eta)\right]$ and $\sin\left[\theta(\xi,\eta)\right]$ are changing much faster than point spread function $h(x,y;\xi,\eta)$, or, in other words, many uncorrelated values of $\exp\left[i\theta(\xi,\eta)\right]$ are observed within the aperture of the imaging system, the central limit theorem of the probability theory can be used to assert that $b^{re}$ and $b^{im}$ are normally distributed with zero mean and $CF_{c\theta}\left(\vec{i}_1,\vec{i}_2\right)$ and $CF_{s\theta}\left(\vec{i}_1,\vec{i}_2\right)$ can be regarded, with respect to integration with $h(x,y;\xi,\eta)$, delta function:

$$CF_{c\theta}\left(\vec{i}_1,\vec{i}_2\right)=CF_{s\theta}\left(\vec{i}_1,\vec{i}_2\right)=AV_\theta\left\{\cos^2\left[\theta(\xi,\eta)\right]\right\}\delta\left(\vec{i}_1-\vec{i}_2\right)=$$
$$AV_\theta\left\{\sin^2\left[\theta(\xi,\eta)\right]\right\}\delta\left(\vec{i}_1-\vec{i}_2\right)=\frac{1}{2}\delta\left(\vec{i}_1-\vec{i}_2\right). \qquad (9.1.24)$$

This implies that

$$CF_{b^{re}}\left(\vec{x}_1,\vec{x}_2\right)=\frac{1}{2}A_0^2\int_{-\infty}^{\infty}h\left(\vec{x}_1;\vec{i}\right)h^*\left(\vec{x}_2;\vec{i}\right)d\vec{i}. \qquad (9.1.25,\text{ a})$$

In the same way one can obtain that:

$$CF_{b^{im}}\left(\vec{x}_1,\vec{x}_2\right)=CF_{b^{re}}\left(\vec{x}_1,\vec{x}_2\right)=\frac{1}{2}A_0^2\int_{-\infty}^{\infty}h\left(\vec{x}_1;\vec{i}\right)h^*\left(\vec{x}_2;\vec{i}\right)d\vec{i}. \qquad (9.1.25,\text{ b})$$

From Eqs. 9.1.24 and 25 it follows that variances of orthogonal components $b^{re}$ and $b^{im}$ are

$$\sigma_b^2=\overline{\left(b^{re}\right)^2}=\overline{\left(b^{im}\right)^2}=\frac{1}{2}A_0^2\int_{-\infty}^{\infty}\int_{-\infty}^{\infty}\left|h\left(x,y;\xi,\eta\right)\right|^2 d\xi\,d\eta. \qquad (9.1.26)$$

In general, they are functions of coordinate $(x,y)$ in the image plane. For space invariant imaging system, when $h(x,y;\xi,\eta)=h(x-\xi,y-\eta)$, they are coordinate independent:

$$\sigma_b^2=\overline{\left(b^{re}\right)^2}=\overline{\left(b^{im}\right)^2}=\frac{1}{2}A_0^2\int_{-\infty}^{\infty}\int_{-\infty}^{\infty}\left|h\left(x,y\right)\right|^2 dx\,dy \qquad (9.1.27)$$

Having asserted that the probability density functions of orthogonal components $b^{re}$ and $b^{im}$ defined by Eqs. 9.1.20 and 21 are normal ones, one can now find the probability density of $I(x,y)$ as the probability density of the sum of squared independent variables $b^{re}$ and $b^{im}$ with normal distribution. To this end, introduce random variables $\rho$ and $\vartheta$ such that

$$b^{re}=\rho\cos\vartheta;\quad b^{im}=\rho\sin\vartheta;\quad I(x,y)=\rho^2. \qquad (9.1.28)$$

Probability that $b^{re}$ and $b^{im}$ take these values within a rectangle ($db^{re}\times db^{im}$) is

$$\frac{1}{2\pi\sigma_b^2}\exp\left[-\frac{\left(b^{re}\right)^2+\left(b^{im}\right)^2}{2\sigma_b^2}\right]db^{re}\,db^{im} =$$

$$\frac{d\vartheta}{2\pi}\frac{\rho}{\sigma_b^2}\exp\left(-\frac{\rho^2}{2\sigma_b^2}\right)d\rho = \frac{d\vartheta}{2\pi}\frac{1}{2\sigma_b^2}\exp\left(-\frac{\rho^2}{2\sigma_b^2}\right)d\rho^2 . \qquad (9.1.29)$$

This is the joint probability of variables $\left\{\rho^2,\vartheta\right\}$. It follows from Eq. 9.1.29 that $\left\{\rho^2 = I(x,y)\right\}$ and $\vartheta$ are statistically independent, $\vartheta$ is uniformly distributed in the range $\left[0,2\pi\right]$ and $I(x,y)$ has the exponential distribution density:

$$P(I) = \frac{1}{2\sigma_b^2}\exp\left(-\frac{I}{2\sigma_b^2}\right), \qquad (9.1.30)$$

where $\sigma_b^2$ is defined by Eq. (9.1.27).

From Eqs. 9.1.19 and 27 one can immediately see that mean value of $I(x,y)$ is:

$$\overline{I(x,y)} = \overline{\left|b^{re}\right|^2} + \overline{\left|b^{im}\right|^2} = 2\sigma_b^2 = A_0^2\int\limits_{-\infty}^{\infty}\int\limits_{-\infty}^{\infty}\left|h(x,y)\right|^2 dx\,dy \qquad (9.1.31)$$

It is the property of the exponential distribution that its standard deviation is equal to its mean value. Therefore

$$\sigma_I = \overline{I(x,y)} = A_0^2\int\limits_{-\infty}^{\infty}\int\limits_{-\infty}^{\infty}\left|h(x,y)\right|^2 dx\,dy . \qquad (9.1.32)$$

Fluctuations of $I(x,y)$ around its mean value form what we call the speckle noise. It is customary to characterize intensity of the speckle noise by ratio of its standard deviation to mean value called "**speckle contrast**":

$$Spckl\_contrast = \frac{\sigma_I}{\overline{I(x,y)}} . \qquad (9.1.33)$$

It follows from Eqs. 9.1.31 and 32 that, for objects, which scatter radiation almost uniformly in the space, speckle contrast of their images obtained in coherent imaging systems that are incapable of resolving microscopic spatial variations in the phase of the object wave front, is unity:

$$Spckl\_contrast = 1. \qquad (9.1.34)$$

Such speckle noise can be regarded as multiplicative with respect to image intensity mean value. Fig. 9.6 illustrates appearance of speckle noise and its spatial correlations that depend on the point spread function of the imaging system (Eqs. 9.1.25). For holographic imaging systems, in which images are reconstructed from holograms, correlation function of the speckle noise depends of the size and shape of the hologram. If images of diffuse objects are reconstructed from smaller and smaller fragments of a hologram, speckle noise in the reconstructed images is becoming more and more correlated thus destroying and hiding small image details, as one can see from images presented in Fig. 9.5, though speckle contrast remains to be equal to unity and images are reconstructed as a whole. This is the explanation of the legendary property of holograms to keep reconstructing images even if the holograms are broken into pieces.

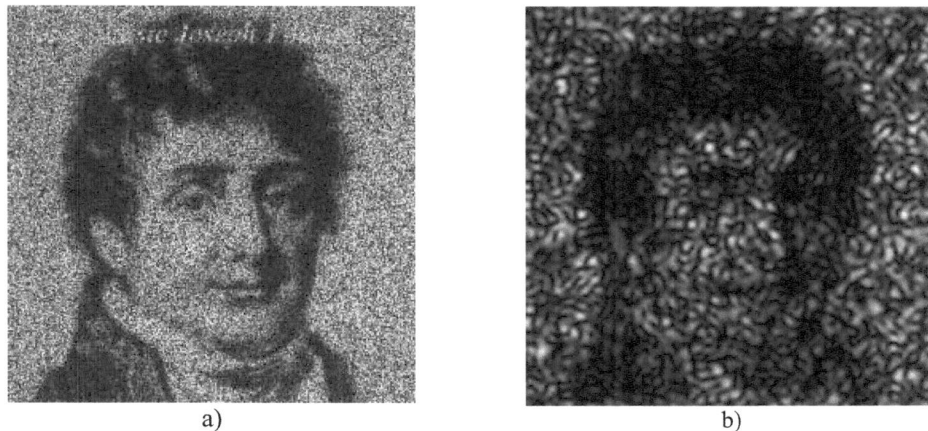

a)　　　　　b)

**Fig. 9.6.** Images with speckle noise for imaging systems with high (a) and low (b) resolving power

The conclusion that speckle contrast for speckle noise, which originates from insufficient resolving power of the observation system is equal to unity is correct only in the assumption that the correlation function of the phase distribution $\exp\left[i\theta(\xi,\eta)\right]$ of the object wave front is much narrower than the point spread function of the imaging system $h(x,y;\xi,\eta)$. In general, speckle contrast depends on the relationship between the width of the correlation function of wavefront phase distribution and the width of the imaging system point spread function, or, in terms of holograms, on the ratio of the size of entire hologram of the object to the size of the part of the hologram used for observing the object. The closer is this ratio to zero, the closer speckle-contrast to its limit value of one. This relationship is illustrated the plots in Fig. 9.7 obtained from numerical simulation results ([1,2]). As one can see from the plots, if this ratio is larger than 0.5, speckle contrast might be significantly lower than unity.

In the above discussion, we considered speckle noise originated from wave front distortions caused by limiting the resolving power of the imaging system. In general, speckle noise in coherent imaging systems appears not only owing to insufficient resolving power of the systems. In fact, any distortions of the wave field that may happen in the wave front sensor such as limitation of the signal dynamic range or signal quantization in the process of its conversion into a digital signal for subsequent storage and processing also cause appearance of speckle noise. Images presented in Fig. 9.8 a)-d) illustrate this phenomenon. As analytical treatment of this problem faces mathematical difficulties, its study can be performed using methods of numerical statistical simulation ([2]).

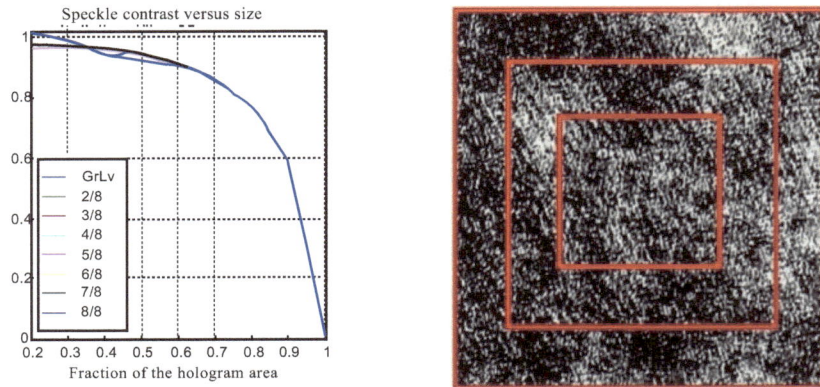

**Fig. 9.7.** Speckle contrast (left) as a function of the fraction of the hologram used for object reconstruction (right); color curves in the plot correspond to different gray level of the image of uniformly painted object; red boxes in the right image outline different fragments of the hologram used for image reconstruction

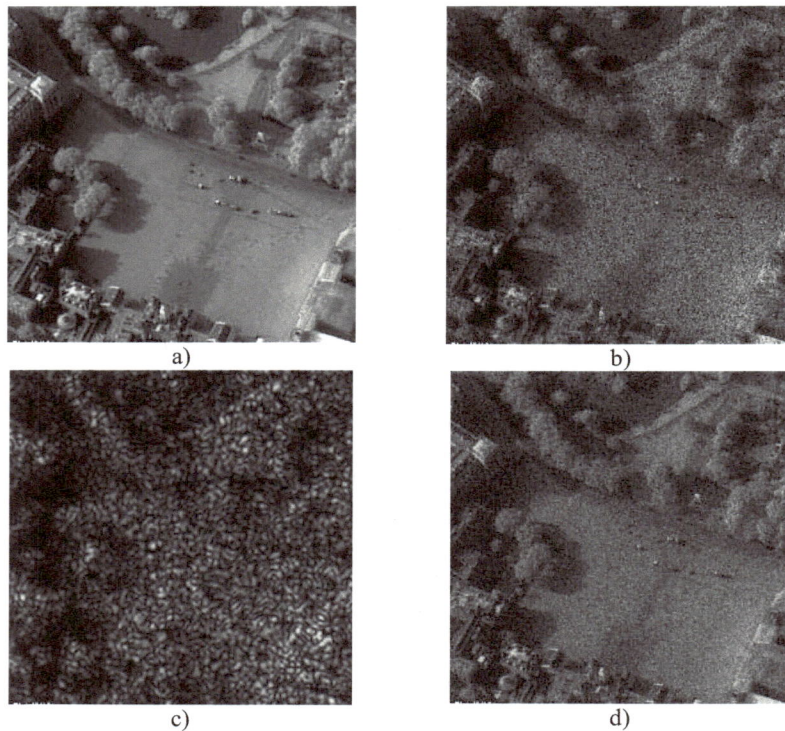

**Fig. 9.8.** Illustrative examples of simulated images: a) - original image; b) - image reconstructed in far diffraction zone from 0.9 of the area of the wave front; c) - image reconstructed in far diffraction zone from 0.5 of the area of the wave front; d) - image reconstructed in far diffraction zone after limitation of the wave front real and imaginary components in the range± their standard deviation.

## 9.2   Measuring parameters of random interferences in sensor and imaging systems

One of the tasks in processing images, holograms and interferograms is removal random interferences caused by imaging, holographic or interferometric sensoric systems. For this, one usually needs to know statistical parameters of the interferences. Sometimes manufacturers of the systems provide the required data. In practice, however, such data are often lacking, and parameters of signal interferences need to be derived directly from image or holographic data under processing.

At the first sight, this problem may seem to be internally contradictory, because, in order to evaluate parameters of noise from noisy data, one must separate the noise from the signal, which can be done only if the noise parameters are known. A way out of this "vicious circle" is not in separating the signal from noise in order to obtain statistical characteristics of the latter, but, rather, in separating signal and noise statistical characteristics in the result of measuring the corresponding characteristics of the noisy signal.

Luckily, in most practical cases random interferences, statistically speaking, are very simple objects; that is, they are described by a very small number of parameters compared to the amount of available data. For instance, additive signal independent white Gaussian noise is fully specified by only two parameters: by its mean and variance and these two parameters are to be estimated from analysis of hundreds of thousands pixels of noisy images. Because of this, noise parameter estimation problem may often be solved by rather simple means. The basic principle for data analysis is choosing those signal characteristics in which the presence of by noise manifests itself in an abnormal behavior of the characteristic that can be detected in the easiest way.

All methods for detection anomalies are based on an a priori assumption of certain smoothness and regularity of the characteristics of undistorted signal that are violated when signal contains noise. Two practical anomaly detection methods are the prediction method, and the voting method.

The prediction method requires that, for every element of the analyzed sequence, the difference between its actual value and the value predicted by previously investigated elements be found. If this difference exceeds certain given threshold, a decision is made that an abnormal burst has been encountered. The number of previously investigated elements (prediction depth), the method of obtaining a predicted value and the threshold value are parameters that have to be specified a priori. The simplest prediction technique regards measured data samples as a sequence, analyzes it one by one and employs the value of the previously checked sequence element as a predicted value of the each next element. Prediction can also be carried out by means of averaging several previous data samples taken with a certain weights. Optimal values of the weight coefficients can be found from one or another model built empirically from the signal data base.

The essence of the voting method is in considering each element of the analyzed data sequence together with certain number of its neighbors. This set of elements is ordered from its minimal to maximal value and it is checked whether or not the value of the given element belongs to a certain predefined number of extreme (maximal or minimal) value of the ordered set. If the answer is yes, then a decision on the presence of an abnormally high (or, alternatively, abnormally low) value in the given element is made. The number of neighbors and the detection threshold are to be specified a priori from the study of the "normal" behavior of the selected characteristic of the undistorted signal.

We will illustrate these approaches using examples of parameter estimation of additive broadband and narrowband noise.

### 9.2.1 Estimation of variance of additive signal-independent broadband noise in images.

Additive signal-independent noise with normal distribution is fully specified by its variance and auto correlation function. If noise is uncorrelated or weakly correlated (broad band), as is often the case, then its variance and correlation function can be found through the use of the following simple algorithm based on measuring abnormalities in the covariance function of the observed image.

Let it be known that observed signal $\{b_k\}$ of $N$ samples ($k=0,...,N-1$) is contaminated by adding to a "pure", or noise-free signal $\{a_k\}$, zero mean signal independent noise $\{n_k\}$:

$$b_k = a_k + n_k . \tag{9.2.1}$$

Compute the empirical autocorrelation function of signal $\{b_k\}$:

$$R_b(m) = \frac{1}{N}\sum_{k=0}^{N-1} b_k b_{k+m}^* = \frac{1}{N}\sum_{k=0}^{N-1} a_k a_{k+m}^* + \frac{1}{N}\sum_{k=0}^{N-1} n_k n_{k+m}^* +$$

$$\frac{1}{N}\sum_{k=0}^{N-1} a_k n_{k+m}^* + \frac{1}{N}\sum_{k=0}^{N-1} a_k^* n_{k+m} = R_a(r) + R_n(m) + \varepsilon_n , \tag{9.2.2}$$

where

$$R_a(m) = \frac{1}{N}\sum_{k=0}^{N-1} a_k a_{k+m}^* , \tag{9.2.3, a}$$

$$R_n(m) = \frac{1}{N}\sum_{k=0}^{N-1} n_k n_{k+m}^* \tag{9.2.3, b}$$

and

$$\varepsilon_n = \frac{1}{N}\sum_{k=0}^{N-1} a_k n_{k+m}^* + \frac{1}{N}\sum_{k=0}^{N-1} a_k n_{k+m}^* \tag{9.2.3, c}$$

One can see from Eqs. 9.2.3 that, owing to the noise additivity, the empirical autocorrelation function of the observed signal $R_b(m)$ is a sum of the empirical autocorrelation function the noise-less signal $R_a(m)$, the empirical autocorrelation function $R_n(m)$ of the noise and of a realization of some "random" process $\varepsilon_n$. The variance of errors in estimating correlation functions through empirical ones measured over finite numbers of signal samples as well as of deviations of the random process $\varepsilon_n$ from zero mean is known to be inversely proportional to the number of samples $N$ of the realization used for the measurement. Normally, in image processing, hundreds of thousands of pixels can be involved in the measurement. Therefore, the variance of estimation errors in Eqs. (9.2.3) is usually sufficiently small, and noise correlation function $\bar{R}_n(m)$ can be, with a good accuracy, estimated through the empirical correlation function as:

$$\bar{R}_n(m) \cong R_n(m) = R_b(m) - R_a(m). \tag{9.2.4}$$

Consider, first, the uncorrelated noise case, when

$$\bar{R}_n(m) = \sigma_n^2 \delta(m) = \begin{cases} \sigma_n^2, m = 0 \\ 0, \ m > 0 \end{cases}, \tag{9.2.5}$$

where $\sigma_n^2$ is the noise variance. In this case, autocorrelation function of the observed image deviates from autocorrelation function of the noise-less image only at $m = 0$ and the difference is equal to the noise variance:

$$\sigma_n^2 \cong R_b(0) - R_a(0). \tag{9.2.6}$$

For all other values $m$, the magnitude of $R_a(m)$ can serve as an estimate of $R_b(m)$. Assuming that $R_a(m)$ is a monotonic function in the vicinity of $m = 0$, one can use $R_a(m) = R_b(m)$ for $m > 0$ to extrapolate its value $\bar{R}_a(0)$ for $r = 0$ and then use the extrapolated value to compute noise variance from Eq. 9.2.6.

A similar approach can be used to estimate variance of a weakly correlated (broad band) noise, i.e. a noise whose autocorrelation function $R_n(m)$ is nonzero only in a small vicinity around the coordinate origin. In this vicinity, the values of the noise-less image correlation function can be quite satisfactorily obtained by extrapolating $R_a(m)$ from points where $R_n(m)$ is known to be zero. Fig. 9.9 illustrates this idea on an example of measuring noise in an interferogram.

### 9.2.2 Estimating intensities and frequencies of spectral components of periodic and similar narrow-band noise.

Additive periodic (moiré) interferences are very characteristic for electronic imaging systems and image sensors. Typical for this type of interference is a feature that their Fourier spectrum contains just a few isolated components. On the other hand, the spatial spectra of noise-free images are generally quite smooth functions more or less monotonically decaying toward to high frequencies. Therefore, the presence of a narrow-band noise in images manifests itself in the form of abnormally large and localized peaks in image power spectra and it is image power spectrum that has to be analyzed for the noise diagnostics.

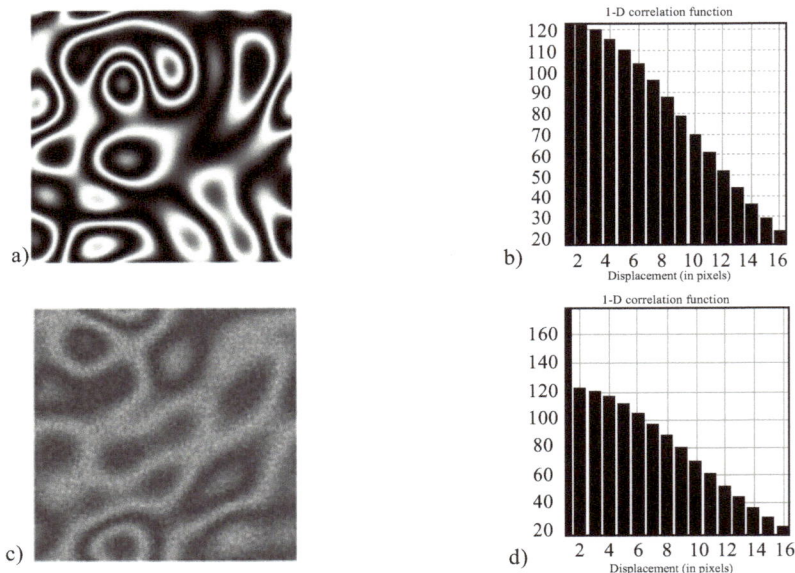

**Fig. 9.9.** Measurement noise level in interferograms: a) – a noise-less interferogram; b) - its empirical 1-D autocorrelation function obtained by averaging of empirical autocorrelation function of image rows; c) – a noisy interferogram; noise variance 10000 (in certain units); d) - 1-D autocorrelation function of the noisy interferogram. One can clearly see an anomalous peak in the correlation function of the noisy interferogram in the origin of coordinate.

In contrast to the case considered above in which the broad band noise produces anomalous peaks in the correlation function only in the vicinity of the coordinate origin, the locations of peaks of noise spectrum in spectral domain are usually unknown and have to be determined. This can be achieved with the help of the above-outlined prediction or voting techniques outlined earlier.

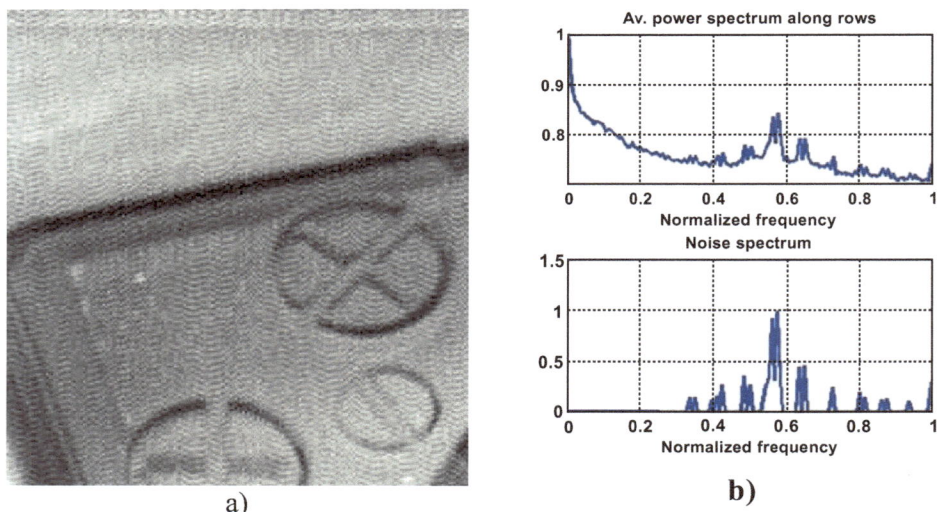

**Fig. 9.10.** Diagnostics of periodical interferences: a) - noisy image; b) column-wise averaged power spectrum of the noisy image (top) and the result of estimation of spectrum of the interference (bottom)

We will explain the diagnostics of the periodical noise in images using as an example an image shown in Fig. 9.10, a). As one can see, the image is distorted by a periodical interference. The top graph of Fig. 9.10, b) shows 1-D power spectrum obtained by computing squared absolute values of DFT spectra of image columns and summing up spectra of all columns. One can easily see anomalous outstanding peaks in image high frequencies that can be attributed to the interference. In order to detect this peak and mark spectral components affected by the interference, one can analyze this spectrum beginning from low frequencies toward higher frequencies employing the prediction or voting techniques.

Since noise is additive, the intensity of each distorted spectral component $P_b(r)$ is equal to the sum of the intensity $P_n(r)$ of the corresponding spectral component of noise and of that $P_a(r)$ of noise-free signal. The latter can be estimated by interpolation between those image spectral components that are found not affected by the interference assuming smoothness of noise-free image power spectrum. Power spectrum of the interference can then be found as:

$$P_n(m) = P_b(m) - \hat{P}_a(m).$$
(9.2.40)

This is illustrated by the bottom graph in Fig. 9.10, b),

The same approach with analysis of image power spectra can be employed for diagnostics of additive broad band noise in interferograms with a spatial carrier, illustrated by an example shown in Fig. 9.11, a). In spectra of noisy interferograms, noiseless interferograms will manifest themselves in concentrated peaks in spectral domain (see Fig. 9.11, b)) that can be easily detected and marked. Then noise variance can be straightforwardly evaluated over the area in spectral domain outside the marked signal peaks.

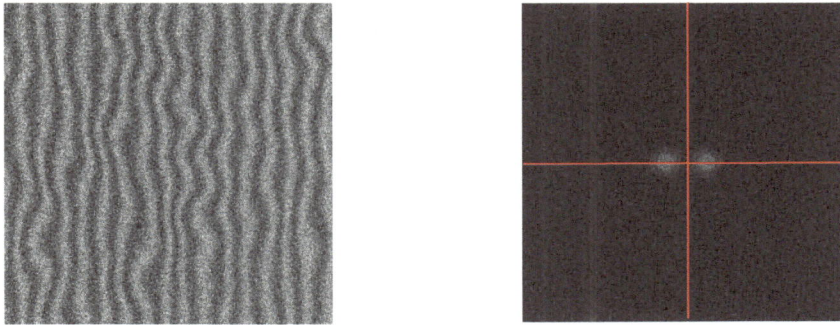

**Fig. 9.11.** Noisy interferogram with a spatial carrier (left) and its power spectrum displayed as an image (right). Vertical and horizontal directions in this image correspond to vertical and horizontal frequencies, respectively while brightness represents spectrum intensity.

## 9.3   Image perfecting methods

Image perfecting is processing aimed at correcting image signal distortions introduced by imaging systems. In this Section, we will review methods of noise reduction and methods of correction of distortions associated with finite apertures of image sensor and display devices and hologram recording devices and distortions caused by nonlinearities and quantization in imaging and holographic devices.

### 9.3.1    *Transform domain MSE optimal scalar Wiener filters*

Consider image restoration for a reduced imaging system model of Fig. 9.12 that disregards point-wise nonlinear transformation in imaging systems and treats image distortions as a combination of only distortions caused by linear filtering plus those caused by action of stochastic interferences.

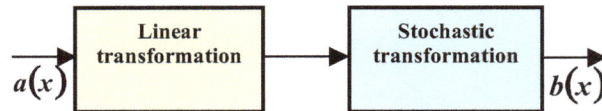

**Fig. 9.12.** A reduced model of linear imaging systems.

The design of the optimal restoration procedure requires specifying a criterion for evaluating the image restoration quality. Let $\{b_k\}$ be a set of $N$ signal samples ($k = 0,1,...,N-1$) at the output of the imaging system, $\{a_k\}$ be a set signal samples that model a perfect, or "ideal" signal and $\{\hat{a}_k\}$ be a set of restored signal samples produced out of samples $\{b_k\}$ by the image restoration processing . For the sake of generality, we will consider the set $\{a_k\}$ as a realization taken from a statistical ensemble $\Omega_A$ or data base of perfect signals and the set $\{b_k\}$ as a realization of a signal ensemble generated by ensembles $\Omega_A$ and $\Omega_N$ of perfect signals and of random interferences, correspondingly.

Define the restoration procedure performance measure as a squared difference between restored and perfect signals averaged over the available set of signal samples and over statistical ensembles $\Omega_A$ and $\Omega_N$. This measure is called ***mean squared restoration error*** (MSE).  The restoration procedure $\{b_k\} \Rightarrow \{\hat{a}_k\}$:

$$\{\hat{a}_k\} = \underset{\mathrm{R}\{b_k\}=\{\hat{a}_k\}}{\arg\min}\left\{\mathrm{AV}_{\Omega_A}\mathrm{AV}_{\Omega_N}\left(\sum_{k=0}^{N-1}|a_k - \hat{a}_k|^2\right)\right\}. \tag{9.3.1}$$

that minimizes this difference will be referred to as ***MSE-optimal filtering***.

For the implementation of the MSE-optimal filtering, we will restrict ourselves to linear filtering, although, in principle, nonlinear filtering can also be used ([1]). In general, linear filtering of a discrete signal can be described in a matrix form as multiplication of a vector of input signal samples $\mathbf{B} = \{b_k\}$ by a filter matrix $\mathbf{H}$:

$$\hat{\mathbf{A}} = \mathbf{H} \cdot \mathbf{B}, \tag{9.3.2}$$

where $\hat{\mathbf{A}} = \left\{ \hat{a}_k \right\}$ is a vector of filter output signal samples

For signals of $N$ samples, a general vector filter matrix $\mathbf{H}$ has dimensions $N \times N$. Specification of such a filter requires determining $N^2$ filter coefficients, and the filtering itself requires performing $O(N^2)$, multiplication and addition operations per $N$ signal samples, or $O(N)$ operation per sample. In image processing, the computational complexity of both determination of filter coefficients and of the filtering maight become too high because of high dimensions of image arrays. Luckily, fast orthogonal transforms, such as Fast Fourier transform and alike, are available that can be computed with the complexity of the order of $O(\log N)$ operations per sample, which allows to radically decrease the filter design and the implementation complexities. Therefore, in what follows, we will consider only filtering in a domain of orthogonal transforms that can be computed with fast algorithms. This class of filters is described by the equation:

$$\hat{\mathbf{A}} = \hat{\mathbf{O}}^{-1} \cdot \mathbf{\varsigma_d} \cdot \hat{\mathbf{O}} \cdot \hat{\mathbf{A}}, \tag{9.3.3}$$

where $\hat{\mathbf{O}}$ and $\hat{\mathbf{O}}^{-1}$ are, correspondingly, direct and inverse matrices of an orthogonal transform that features fast transform algorithm, $\mathbf{\varsigma} = \mathbf{diag}\left\{ \eta_r \right\}$, ($r = 0, 1, ..., N-1$) is a diagonal filter matrix of filter coefficients $\left\{ \eta_r \right\}$ in the transform domain. Such filtering is called ***"scalar" filtering***. The scalar filtering implies the following relationship between transform coefficients $\left\{ \hat{\alpha}_r \right\} = \hat{\mathbf{O}} \cdot \hat{\mathbf{A}}$ and $\left\{ \beta_r \right\} = \hat{\mathbf{O}} \cdot \mathbf{B}$ of output and input signal samples:

$$\hat{\alpha}_r = \eta_r \beta_r. \tag{9.3.4}$$

In the assumption of the orthogonality of the transform $\hat{\mathbf{O}}$, one can, by virtue of the Parceval relationship, modify the filter optimality condition defined by Eq. 9.3.1 in the following way:

$$\left\{ \hat{\alpha}_r \right\} = \underset{\{\eta_r\}}{\arg\min} \left\{ AV_{\Omega_A} AV_{\Omega_N} \left( \sum_{r=0}^{N-1} \left| \alpha_r - \hat{\alpha}_r \right|^2 \right) \right\} = \underset{\{\eta_r\}}{\arg\min} \left\{ AV_{\Omega_A} AV_{\Omega_N} \left( \sum_{r=0}^{N-1} \left| \alpha_r - \eta_r \beta_r \right|^2 \right) \right\}. \tag{9.3.5}$$

By computing derivatives over sought variables $\left\{ \eta_r \right\}$ and equaling them to zero, one can obtain from Eq. 9.3.5 that optimal scalar filter coefficients $\left\{ \eta_r^{opt} \right\}$ should be cross-correlation coefficients between spectral coefficients $\beta_r$ and $\alpha_r$ of the filter input and perfect signals ([1]):

$$\eta_r^{opt} = \frac{AV_{\Omega_A} AV_{\Omega_N} \left( \alpha_r \beta_r^* \right)}{AV_{\Omega_A} AV_{\Omega_N} \left( \left| \beta_r \right|^2 \right)}. \tag{9.3.6}$$

MSE optimal linear filters and, in particular, MSE optimal scalar filters defined by Eq. 9.3.6 are called *scalar Wiener filters*.

In order to implement a scalar Wiener filter, one should therefore know cross-correlation $\left\{ \text{AV}_{\Omega_A} \text{AV}_{\Omega_N} \left( \alpha_r \beta_r^* \right) \right\}$ between filter input signal and perfect signal spectral coefficients and also power spectrum $\left\{ \text{AV}_{\Omega_A} \text{AV}_{\Omega_N} \left( \left| \beta_r \right|^2 \right) \right\}$ of the input signal in the selected basis. The statistical approach we adopted assumes averaging of the restoration error over statistical ensembles of perfect signals and of filter input signals. This implies that the required statistical parameters should be measured in advance for these ensembles or over the databases.

### 9.3.2    *Filtering additive signal independent noise and empirical Wiener filters*

Consider a noise model, in which filter input signal samples $\left\{ b_k \right\}$ are obtained as a sum of perfect signal samples $\left\{ a_k \right\}$ and samples $\left\{ n_k \right\}$ of signal independent zero mean random noise:

$$b_k = a_k + n_k .$$

$$(9.3.7)$$

In spectral domain, the same relationship holds for signal and noise spectral coefficients:

$$\beta_r = \alpha_r + v_r ,$$

$$(9.3.8)$$

where $\left\{ v_r \right\} = \mathbf{T} \left\{ n_k \right\}$. For this model one can obtain that

$$\text{AV}_{\Omega_A} \text{AV}_{\Omega_N} \left( \alpha_r \beta_r^* \right) = \text{AV}_{\Omega_A} \text{AV}_{\Omega_N} \left[ \alpha_r \left( \alpha_r^* + v_r^* \right) \right] = \text{AV}_{\Omega_A} \left( \left| \alpha_r \right|^2 \right) \qquad (9.3.9)$$

and

$$\text{AV}_{\Omega_A} \text{AV}_{\Omega_N} \left( \left| \beta_r \right|^2 \right) = \text{AV}_{\Omega_A} \text{AV}_{\Omega_N} \left[ \left( \alpha_r + v_r \right) \left( \alpha_r^* + v_r^* \right) \right] = \text{AV}_{\Omega_A} \left( \left| \alpha_r \right|^2 \right) + \text{AV}_{\Omega_N} \left( \left| v_r \right|^2 \right)$$

$$(9.3.10)$$

The latter is because of zero mean noise: $\text{AV}_{\Omega_N} \left( v_r^* \right) = \text{AV}_{\Omega_N} \left( v_r \right) = 0$. Therefore the scalar Wiener filters for suppressing additive signal independent noise are defined through their transform domain coefficients $\left\{ \eta_r \right\}$ as

$$\eta_r^{opt} = \frac{\text{AV}_{\Omega_A} \left( \left| \alpha_r \right|^2 \right)}{\text{AV}_{\Omega_A} \left( \left| \alpha_r \right|^2 \right) + \text{AV}_{\Omega_N} \left( \left| v_r \right|^2 \right)} .$$

$$(9.3.11)$$

This formula has a clear physical interpretation. Define ***signal-to-noise ratio*** for $r$-th transform coefficient as:

$$SNR_r = \frac{AV_{\Omega_A}\left(|\alpha_r|^2\right)}{AV_{\Omega_N}\left(|v_r|^2\right)}.$$ 

(9.3.12)

Then obtain:

$$\eta_r^{opt} = \frac{SNR_r}{1+SNR_r},$$ 

(9.3.13)

which means that scalar Wiener filter weight coefficients are defined, for each signal spectral coefficient, by the signal-to-noise ratio for this coefficient. The lower is signal-to-noise ratio for a particular signal spectral component, the lower the contribution of this component to the filter output signal.

In order to implement a scalar Wiener filter one has to know in advance power spectra $AV_{\Omega_A}\left(|\alpha_r|^2\right)$ and $AV_{\Omega_N}\left(|v_r|^2\right)$ of perfect signals and of noise in the selected transform domain. Noise power spectrum may be known from the specification certificate of the imaging device. Otherwise it can be measured in noisy input signals using methods described in Sect. 9.2. As for the perfect signal power spectrum, it is most frequently not known. However, one can, using Eq. 9.3.10, attempt to estimate it from the power spectrum $AV_{\Omega_A}AV_{\Omega_N}\left(|\beta_r|^2\right)$ of input noisy signals as

$$AV_{\Omega_A}\left(|\alpha_r|^2\right) = AV_{\Omega_A}AV_{\Omega_N}\left(|\beta_r|^2\right) - AV_{\Omega_A}\left(|v_r|^2\right).$$ 

(9.3.14)

The latter has to be estimated from the observed signal spectrum $|\beta_r|^2$. Denote this estimate $\overline{|\beta_r|^2}$. This empirical estimate made by averaging over available realizations of input images may, when used as a replacement for $AV_{\Omega_A}AV_{\Omega_N}\left(|\beta_r|^2\right)$, give negative values for some transform coefficients because of the limited depth of the averaging. Since power spectra cannot assume negative values, the following modified spectrum estimation can be adopted:

$$AV_{\Omega_A}\left(|\alpha_r|^2\right) = \max\left[\overline{|\beta_r|^2} - AV_{\Omega_A}\left(|v_r|^2\right);\ 0\right].$$ 

(9.3.15)

In this way we arrive at the filter:

$$\eta_r = \max\left[ \frac{\overline{|\beta_r|^2} - \mathrm{AV}_{\Omega_\Lambda}\left(|\nu_r|^2\right)}{\overline{|\beta_r|^2}}; \quad 0 \right].$$
(9.3.16)

We will refer to this filter as the ***empirical Wiener filter***.

If the imaging system noise is known to be non-correlated (white) noise with variance $\sigma_n^2$, the empirical Wiener filter coefficients are computed as:

$$\eta_r = \max\left[ \frac{\overline{|\beta_r|^2} - \sigma_n^2}{\overline{|\beta_r|^2}}; \quad 0 \right].$$
(9.3.17)

As a zero order approximation to the input images power spectrum, power spectrum $|\beta_r|^2$ of a single input image subjected to filtering can be used. In this case, empirical Wiener filter weight coefficients are found as:

$$\eta_r = \max\left[ \frac{|\beta_r|^2 - \sigma_n^2}{|\beta_r|^2}; \quad 0 \right].$$
(9.3.18)

Note that such an empirical Wiener filter is adaptive because its weight coefficients depend on the spectrum of the image to which it will be applied.

Weight coefficients of scalar Wiener filters assume values in the range between zero and one. A version of the empirical Wiener filter of Eq. 9.3.18 with binary weight coefficients:

$$\eta_r = \begin{cases} 1, & if \ |\beta_r|^2 \geq Thr \\ 0, & otherwise \end{cases},$$
(9.3.19)

where *Thr* is a rejecting threshold, is called the ***rejecting filter***. As it follows from Eq. 9.3.18, the rejecting threshold has a value of the order of magnitude of noise variance $\sigma_n^2$. Rejecting filters eliminate from input images all components, for which signal-to-noise ratio is lower than a certain threshold commensurable with the noise variance.

Described empirical Wiener filters for signal denoising are particularly very efficient if signal and/or noise spectra are well concentrated and are separated in the transform domain selected for processing. A typical example of such a situation is filtering of quasi-periodical noise interferences, whose spectrum has only a few components in the domain of Discrete Fourier (DFT) or Discrete Cosine (DCT) Transforms. Fig. 9.13 demonstrates the empirical Wiener filtering quasi-periodical noise pattern in an image.

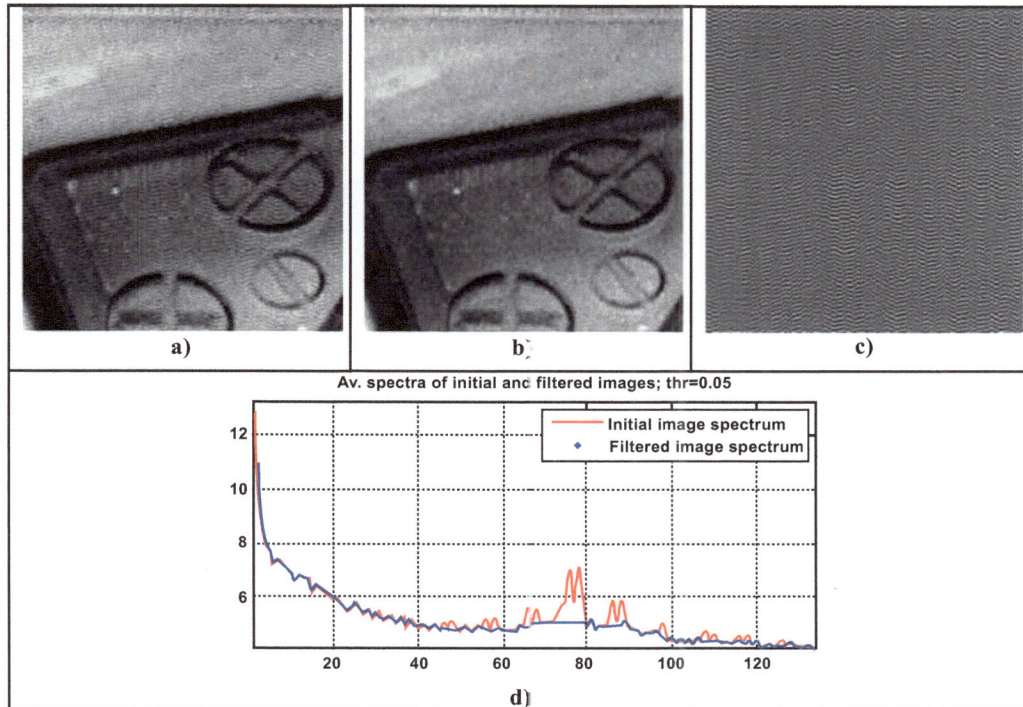

**Fig. 9.13.** Filtering periodic interferences: a) initial noisy image; b) filtered image; c) difference between noisy and filtered images, which shows suppressed noise component; d) DFT spectra of initial and filtered images

One can clearly see anomalous peaks of noise spectrum in the input image spectrum that are eliminated in the output image spectrum after applying empirical Wiener filtering designed using methods of noise spectrum estimation described in Sect. 9. 2. Note that in this particular example filtering was carried out as 1-D row-wise filtering in DFT domain.

Fig. 9.14 illustrates 2-D rejecting filtering for eliminating periodical noise components in a digitized Fresnel hologram before its reconstruction.

The same principle can be applied to filtering wide band noise in interferograms recorded with a spatial carrier. In this case useful signal of the interferogram is very narrow-band in the domain of DFT or DCT, whereas noise spectrum is spread uniformly over entire spectral plane. Upon detecting in Fourier domain peak of the spectrum of the interferogram using methods described in Sect. 9.2, one can synthesize an empirical Wiener or a rejecting filter and then apply it to the noisy interferogram. A result of such processing is illustrated in Fig. 9.15.

In image processing applications, Wiener filtering of wide band noise and especially white noise is much less efficient than in filtering narrow–band noise. When filtering white noise, Wiener filter tends to weaken low energy signal spectral components. However these components are usually exactly the components that are the most important in images because they carry information about object boundaries, which is critical for object detection and recognition. Moreover, Wiener filtering converts input white noise into a correlated output residual noise, though with a reduced variance. As one can see from Eq. 9.3.11, in case of the intensive input white noise, power spectrum of the residual noise is roughly proportional to the signal power spectrum which means that the residual noise becomes, statistically, signal alike. This usually substantially hampers subsequent image visual analysis. As a result, Wiener filtering for image denoising in such case may even worsen processed image visual quality. Note that local adaptive empirical Wiener filters introduced in Sect. 9.3.4 are free this drawback.

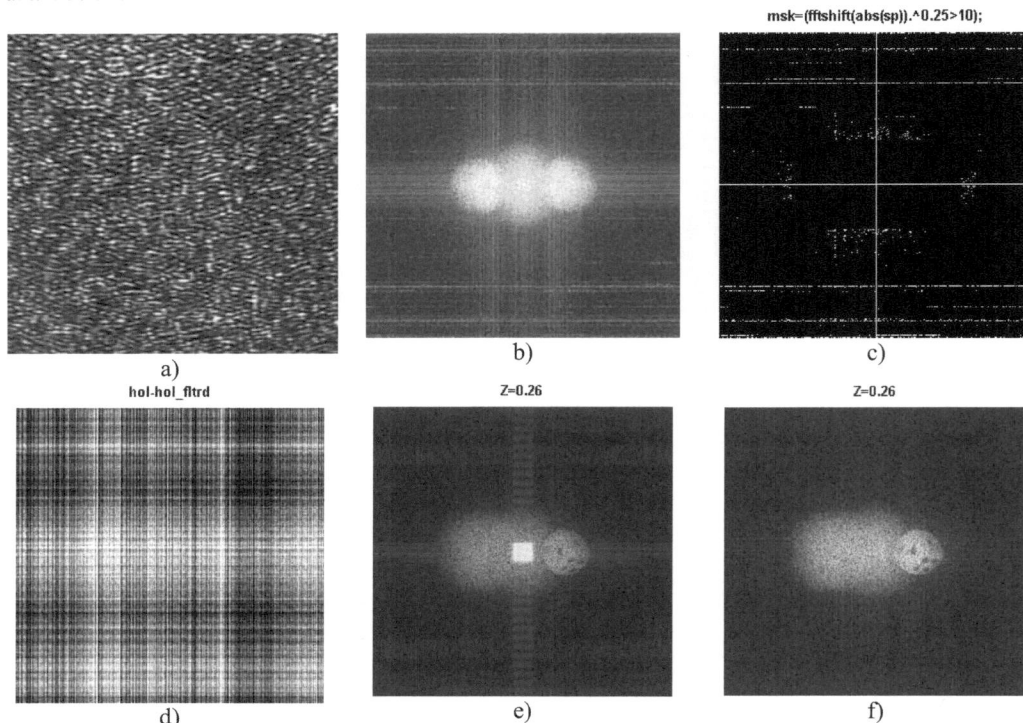

Fig. 9.14. Cleaning quasi-periodic interferences in Fresnel holograms before their numerical reconstruction: a) a digitized hologram; b) DFT spectrum of the hologram, which striation reveals the presence of quasi-periodic interference; c) pattern of 2-D rejecting filter coefficients (black for 1 and white for 0); d) difference between initial hologram and filtered one demonstrating noise pattern removed by the filtering; e) image reconstructed from the initial hologram; f) image reconstructed from the filtered hologram, which shows removal of interferences seen in figure e). (The hologram was kindly granted by Dr. T. Kraus, BIAS, Bremen, Germany)

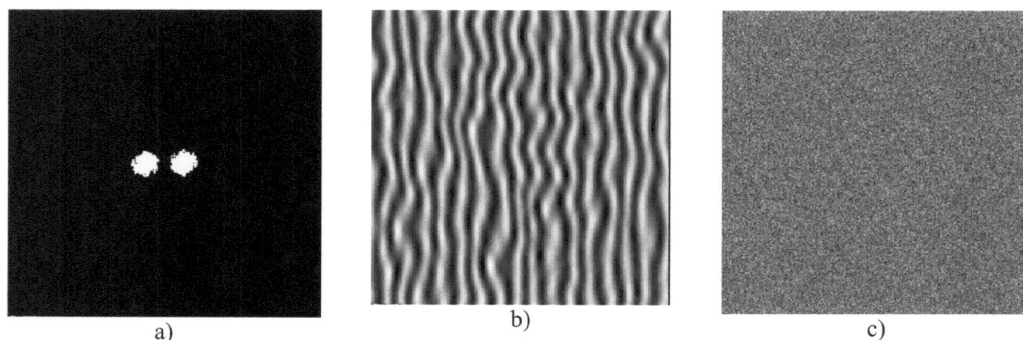

**Fig. 9.15.** Denoising interferogram shown in Fig. 9.5 using a rejecting filter: a) rejecting filter frequency response in DFT domain (white areas are frequencies, where the filter frequency response is one, black areas are frequencies, where the filter frequency response is zero); b) filtering result; c) difference between noisy interferogram and the filtered one, which shows noise removed from the interferogram.

A version of the empirical Wiener filtering implemented with wavelet transform image decomposition is known as ***wavelet shrinkage*** filtering ([4]). Empirical Wiener filtering according to Eq. 9.3.18 is called on the wavelet jargon "***soft thresholding***". Rejecting filtering according to Eq. 9.3.19 is called "***hard thresholding***". Fig. 9.16 shows a flow diagram of signal denoising by the wavelet shrinkage. As one can see from the flow diagram, image filtering is carried in multiple scales obtained recursively. At each scale, the image is decomposed to a low frequency component, obtained by low pass filtering to half of the initial bandwidth, and a residual high frequency component. The low frequency component is sub-sampled (down sampled) to half of the initial sampling rate. This downsampled low pass component is used as an input image for the next scale. After being interpolated to the initial, it is also used for formation the high frequency component. The latter is the component that is subjected, pixel-by-pixel, to soft or hard thresholding according to, correspondingly, Eqs. 9.3.18 and 9.3.19, in which its pixel values play the role of the spectral coefficients.

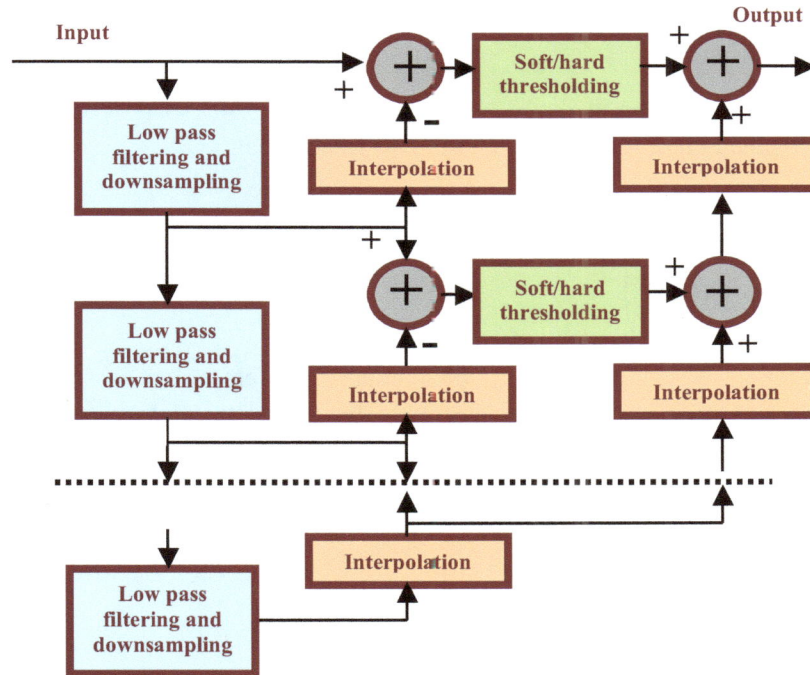

**Fig. 9.16.** Wavelet shrinkage: signal denoising in wavelet transform domain

It is shown in Ref. 3 that working in wavelet transform domain and as well as filtering in sliding window described below in Sect. 9.3.4 can be regarded as different implementations of image sub-band decomposition.

### 9.3.3 *Image deblurring, inverse filters and aperture correction.*

Consider now an imaging system model that accounts also for signal linear transformations in imaging systems. Suppose that the linear transformation unit can be modeled as a scalar filtering with filter coefficients $\{\lambda_r\}$ in the selected transform domain. For DFT and DCT bases, this assumption is justified, to the accuracy of boundary effects, for shift invariant linear filtering. In this case coefficients $\{\lambda_r\}$ are samples of the imaging system frequency response. In ideal imaging systems, they should all be equal to unity. In reality, they decay with the frequency index $r$, which results, in particular, in image blur. Processing aimed at correcting this type of distortions is frequently referred to as *image deblurring*.

For such systems, we have in the transform domain:

$$\beta_r = \lambda_r \alpha_r + v_r, \tag{9.3.20}$$

Using Eqs. 9.3.6, one can obtain that the scalar Wiener image restoration filter is defined in this case by the equation:

$$\eta_r = \frac{1}{\lambda_r} \frac{SNR_r}{1 + SNR_r}, \tag{9.3.21}$$

where $SNR_r$ is signal-to-noise ratio on *r*-th component in the transform domain at the output of the linear filter unit of the imaging system model:

$$SNR_r = \frac{|\lambda_r|^2 \, \mathrm{AV}_{\Omega_A}\left(|\alpha_r|^2\right)}{\mathrm{AV}_{\Omega_N}\left(|v_r|^2\right)}. \tag{9.3.22}$$

Correspondingly, the general empirical Wiener filter, empirical Wiener filter with zero order approximation to perfect signal spectrum and the rejecting filter for aperture correcting and image deblurring will be in this case as follows:

$$\eta_r = \max\left[ \frac{1}{\lambda_r} \frac{\overline{|\beta_r|^2} - \mathrm{AV}_{\Omega_A}\left(|v_r|^2\right)}{\overline{|\beta_r|^2}}; \quad 0 \right]; \tag{9.3.23}$$

$$\eta_r = \max\left[ \frac{1}{\lambda_r} \frac{\overline{|\beta_r|^2} - \sigma_n^2}{\overline{|\beta_r|^2}}; \quad 0 \right] \tag{9.3.24}$$

and

$$\eta_r = \begin{cases} \dfrac{1}{\lambda_r}, & if \; |\beta_r|^2 \geq Thr \\[2mm] 0, & otherwise \end{cases}. \tag{9.3.25}$$

All these filters may be treated as two filters in cascade: the filter with coefficients

$$\eta_r^{inv} = \frac{1}{\lambda_r} \qquad (9.3.26)$$

usually called the ***inverse filter*** and signal denoising filters described by Eqs. 9.3.17 - 19.

Inverse filters compensate weakening signal frequency components in the imaging system while denoising filters prevent from excessive amplification of noise of components with small $SNR_r$ and perform what is called "***regularization***" of inverse filters. As one can see from Eq. 9.3.22, weight coefficients of denoising filters for small $\{\lambda_r\}$ decay faster than weight coefficients of the inverse filter grow.

One of the most immediate applications of inverse filters is correcting distortions caused by a finite size of apertures of image sensors, image sampling devices and image displays. We will refer to this processing as ***aperture correction***.

Let the image sensor and sampling device be an array of light sensitive elements with a square aperture of size $d^{(s)} \times d^{(s)}$ as it is shown in Fig. 9.17.

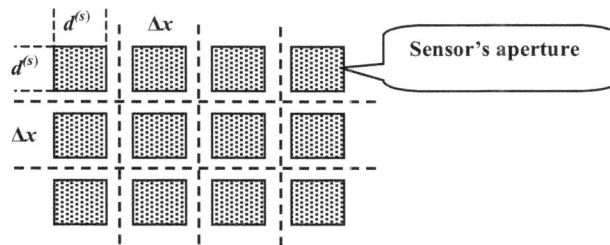

**Fig. 9.17.** Arrangement of light sensitive elements in image sensor arrays

Then frequency response of the sensor is

$$H\left(f_x, f_y\right) = \int\limits_{-d^{(s)}/2}^{d^{(s)}/2} \exp\left(i2\pi f_x x\right) dx \int\limits_{-d^{(s)}/2}^{d^{(s)}/2} \exp\left(i2\pi f_y y\right) dx =$$

$$\frac{\sin\left(\pi f_x d^{(s)}/2\right)}{\pi f_x d^{(s)}/2} \frac{\sin\left(\pi f_y d^{(s)}\right)}{\pi f_y d^{(s)}} = \text{sinc}\left(\pi f_x d^{(s)}\right)\text{sinc}\left(\pi f_y d^{(s)}\right), \qquad (9.3.27)$$

or, in dimensionless coordinates $\left\{\overline{f}_x = f_x \Delta x, \overline{f}_y = f_y \Delta x\right\}$, where $\Delta x$ is the sampling interval in both coordinates,

$$H\left(\overline{f}_x, \overline{f}_y\right) = \text{sinc}\left(\pi \overline{f}_x d^{(s)}/\Delta x\right)\text{sinc}\left(\pi \overline{f}_y d^{(s)}/\Delta x\right). \qquad (9.3.28)$$

The discrete frequency response of the sensor is then:

$$\lambda_{r,s} = \lambda_r \lambda_s = \mathrm{sinc}\left(\pi \overline{d}^{(s)} \lambda_r / N_x\right) \cdot \mathrm{sinc}\left(\pi \overline{d}^{(s)} \lambda_s / N_y\right), \qquad (9.3.29)$$

where $\overline{d}^{(s)} = d^{(s)} / \Delta x$ is the **camera fill-factor**.

If the image display device has also a square aperture of size $d^{(r)} \times d^{(r)}$, the overall imaging system discrete frequency response is:

$$\lambda_{r,s} = \lambda_r \lambda_s = \mathrm{sinc}\left(\pi \overline{d}^{(s)} \lambda_r / N_x\right) \cdot \mathrm{sinc}\left(\pi \overline{d}^{(r)} \lambda_r / N_x\right) \times$$
$$\mathrm{sinc}\left(\pi \overline{d}^{(s)} \lambda_s / N_y\right) \mathrm{sinc}\left(\pi \overline{d}^{(r)} \lambda_s / N_y\right). \qquad (9.3.30)$$

Parameters $d^{(s)}$, $d^{(r)}$ and $\Delta x$ are imaging system design parameters that are usually known from system's certificate. They can be used for correcting image distortions by processing images in computer.

Fig. 9.18 illustrates an example of the aperture correction of an air photograph. Right image in this figure is obtained by applying to the left image an inverse filter for the system's frequency response defined by Eq. 9.3. 20 with $\overline{d}^{(s)} = \overline{d}^{(r)} = 1$.

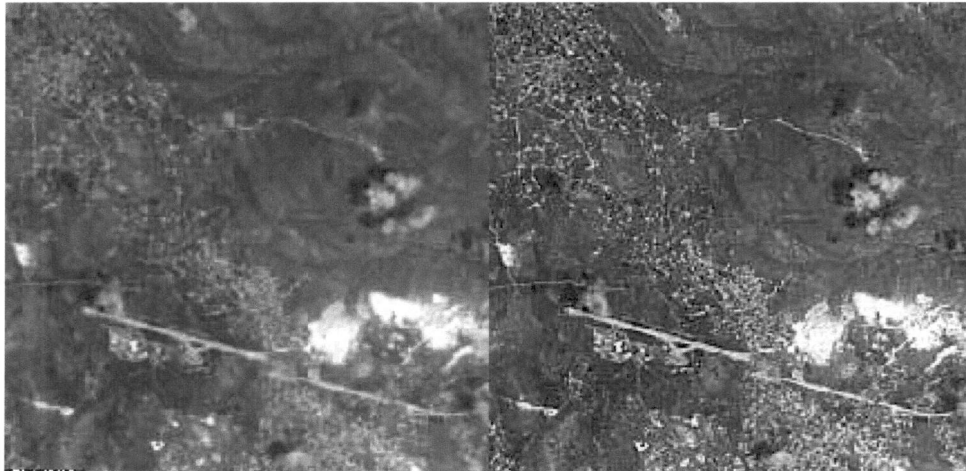

**Fig. 9.18.** Aperture correction: initial (left) and aperture corrected (right) images

In synthesis of computer-generated holograms, aperture correction is required to compensate masking images reconstructed from holograms by the frequency response of the hologram recording device (see Chapt. 7). This masking may result in substantial reducing image contrast on the periphery of image plane. The aperture correction can be implemented by multiplying object image, before the hologram computation, by the function inverse to the hologram recording device frequency response. This method is illustrated in Fig. 9.19.

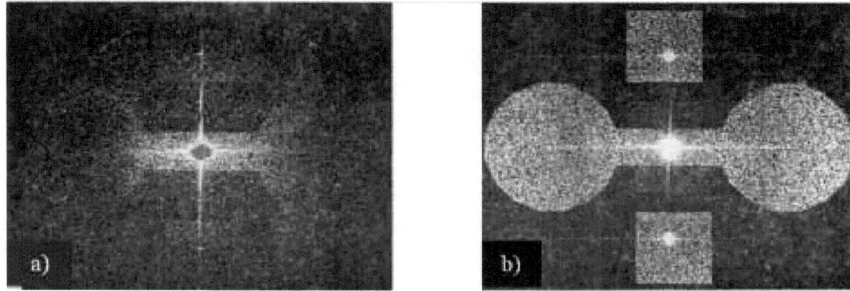

**Fig. 9.19.** Effect of spatial masking of reconstructed images due to the finite size of the hologram recording aperture (a) and a result of its compensation (b) (adopted from [5])

### 9.3.4   *Local adaptive filtering*

In the derivation of the MSE optimal scalar linear filters, the size of input signal vectors was a free parameter. In the filter implementation, one should select the fashion, in which signals are to be processed. Given an image to be processed, filtering can be designed and carried out either over the entire set of available image samples or fragment-wise. We will refer to the former as ***global filtering*** and to the latter as ***local filtering***.

A theoretical framework for local filtering is provided by local criteria of processing quality ([1]). For optimal local MSE scalar filters, filter coefficients $\left\{ \eta_r^{(k)} \right\}$ at $k$-th window position are, according to the local criteria, defined by the equation:

$$\hat{\alpha}_r^{(k)} = \arg\min_{\{\eta_r\}} \left\{ \mathrm{AV}_{\Omega_A} \mathrm{AV}_{\Omega_N} \left( \sum_{r=0}^{W-1} \left| \alpha_r^{(k)} - \eta_r^{(k)} \beta_r^{(k)} \right|^2 \right) \right\}, \qquad (9.3.31)$$

where $W$ is it the filter window size and, correspondingly, the size of the area over which filtering performance is evaluated, $\left\{ \beta_r^{(k)} \right\}$ are spectral coefficients of the filter input samples within the filter window, $\left\{ \alpha_r^{(k)} \right\}$ are spectral coefficients of hypothetical perfect signal samples within the filter window. The solutions obtained above for empirical Wiener filters for image restoration can then be extended to local filtering by replacing in Eqs. 9.3.16 - 19 and 9.3.23-26 signal spectra with corresponding local spectra of the signal window samples.

The most straightforward way to implement local filtering is to perform it in a hopping window. However, "hopping window" processing being very attractive from the computational complexity point of view suffers from "blocking effects" - artifacts in form of discontinuities at the borders of the hopping window. Obviously, the ultimate solution of the "blocking effects" problem would be sliding window processing. Sliding window filtering is local adaptive because, for empirical Wiener filters, filter coefficients depend on local spectra of image fragments.

In the sliding window filtering, one should, for each position of the filter window, compute transform coefficients of the input signal samples within window, design on this base the filter, modify accordingly transform coefficients and then compute inverse transform. With sliding window processing, inverse transform need not, in principle, be computed for all signal samples within the filter window since only the central sample of the window has to be determined in order to form the output signal in the process of image scanning with the filter window. Fig. 9.20 illustrates this process.

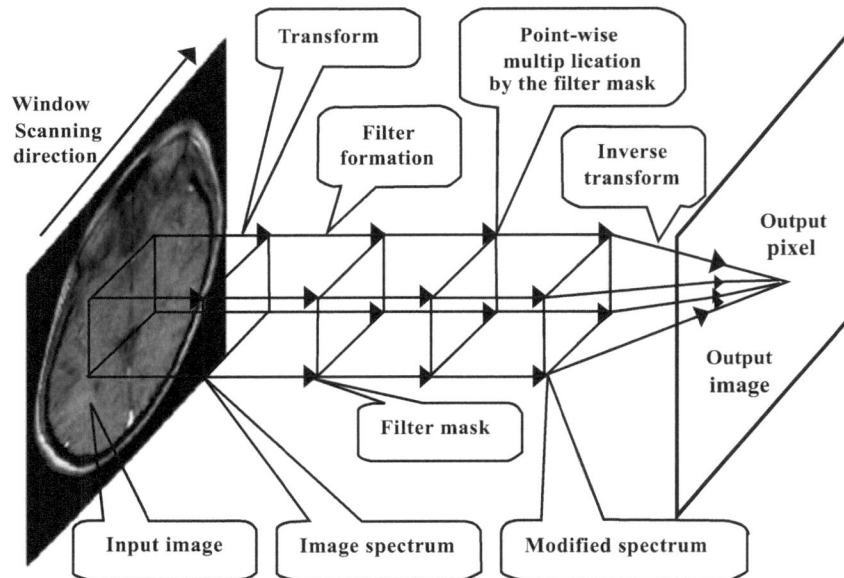

**Fig. 9.20**. The principle of local adaptive filtering in sliding window

Selection of orthogonal transforms for implementation of local adaptive filtering is governed by
- the required accuracy of approximation of general linear filtering with scalar filtering;
- the availability of a priori knowledge regarding image spectra in the chosen transform domain;
- the accuracy of the empirical spectrum estimation from the observed data for the adaptive filter design;
- the computational complexity of the filter implementation.

Among all known transforms, Discrete Cosine Transform proved to be one of the most appropriate transforms for sliding window transform domain filtering ([1,2]). DCT exhibits a very good energy compaction capability, which is a key feature for the efficiency of the filter design and implementation. Being advantageous to DFT in terms of the energy compaction capability, DCT can also be regarded as a good substitute for DFT in signal/image restoration tasks with imaging system specification in terms of their frequency responses. DCT is also quite suitable for multi component image processing. The use of DCT in sliding window has low computational complexity owing to the recursive algorithms for computing DCT in sliding windows

([1]). In addition note that it is advisable to select the window size $N_w$ to be an odd number in order to make the filter window symmetrical with respect to the window central pixel. In this case the inverse DCT transform of local spectrum $\beta_r^{(k)}$ for computing the window central pixel $a_{k+(N_w-)/2}$ at the filter output

$$a_{k+(N_w-1)/2} = \beta_0^{(k)} + 2\sum_{r=1}^{W-1} \beta_r^{(k)} \cos(\pi r/2) = \beta_0^{(k)} + 2\sum_{s=1}^{(W-1)/2} \beta_{2s}^{(k)}(-1)^s \ ,(9.3.32)$$

involves only signal spectrum coefficients with even indices. Therefore only those spectral coefficients have to be computed and the computational complexity of sliding window filtering in DCT domain is $O\left[(N_w+1)/2\right]$ operations for 1-D filtering and $O\left[(N_w+1)^2/4\right]$ operations for 2-D filtering in a square window of $N_w \times N_w$ pixels.

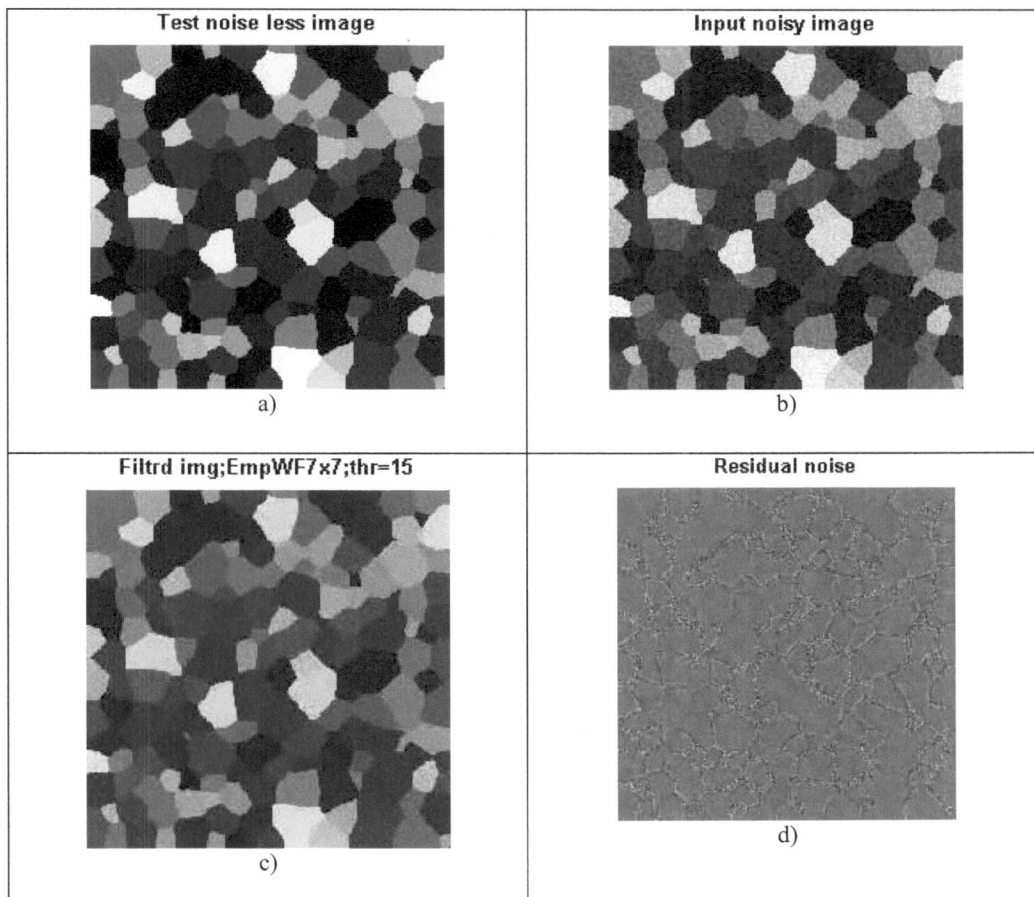

**Fig. 9.21.** Sliding window DCT domain denoising: test image (a) noisy test image (b), filtered image (c) and difference between the initial noiseless filtered images (d).

Fig. 9.21 illustrates local adaptive empirical Wiener filtering in DCT domain of a piece-wise constant test noisy image with noise dynamic range ±20 gray levels of 256. Upper left image is an initial test image. Upper right image is the input image with a noise added. Left bottom image is the result of the filtering. Right bottom image shows residual noise in the filtered image obtained as a difference between noise less image and filtered one. One can see from this image that the residual noise is observed practically only in the vicinity of the borders of piece wise constant patches of the test image. This means that when filter window seats on an image patch, where image gray values are constant, the filter is almost completely opaque and preserves practically only image local dc component (local mean) while on the patch boundaries the filter is almost fully transparent to signal spectral components and prefers not to remove noise in order to preserve edges of patches. This edge preserving capability is an important advantage of the local adaptive filtering compared to the global filtering. It is interesting to note that similar insensitivity to noise in the vicinity of image edges is characteristic for human vision as well.

Local adaptive filtering is capable also of denoising images with signal dependent noise such as speckle noise. The characteristic feature of the speckle noise is that its standard deviation is proportional to the signal (see Sect. 9.1). For filtering speckle noise with local adaptive filters, one should set parameters $\sigma_n$ and *Thr* of the filters of Eqs. 9.3. 17 through 19 to be proportional to the image local mean component $\beta_0^{(k)}$. Fig. 9.22 shows an example of such denoising of an interferogram.

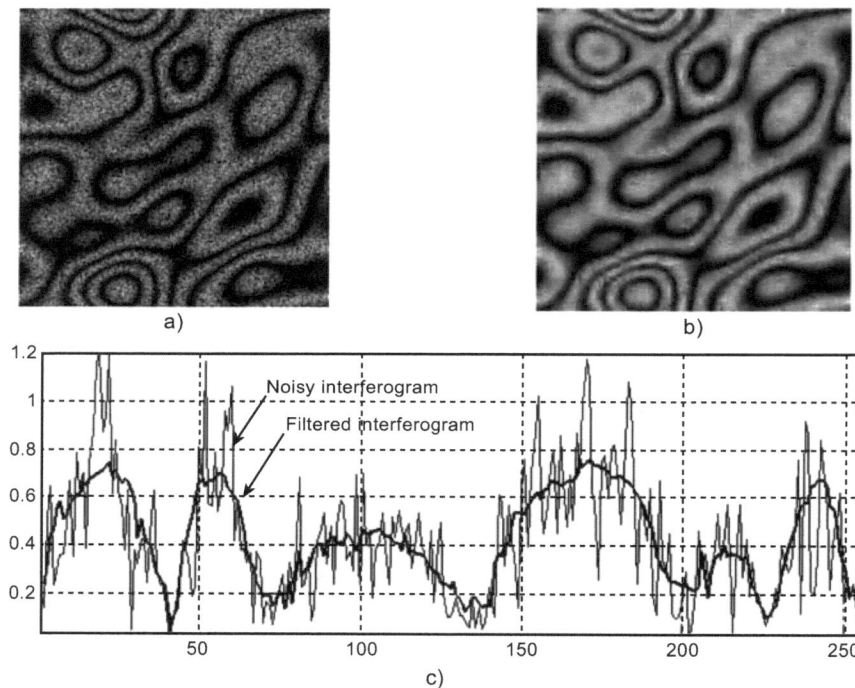

**Fig. 9.22.** An example of local adaptive filtering speckle noise in interferogram: Initial noisy interferogram (a), filtered interferogram (b) and plots of a typical row of noisy and filtered interferograms (c).

## 9.3.5    *Methods for correcting image grey scale non-linear distortions*

In the imaging system model introduced in Sect. 9.1 image gray scale distortions are attributed to the point-wise nonlinear transformation unit. Usually in correcting gray scale distortions image noise component attributed to the stochastic transformations is ignored. In such an assumption, for correcting image nonlinear transformations in imaging systems it is sufficient to apply to images corresponding inverse transformations. However, in digital processing one should take into account quantization effects and the fact that any nonlinear transformation of quantized signals results in reducing the number of signal quantization levels. The phenomenon of slinging quantization levels together as the result of their nonlinear transformation is illustrated in Fig. 9.23.

For generating correcting nonlinear transformations that can be applied to quantized distorted input signal, principles of signal optimal scalar quantization ([1]) should be applied. Assume that signal quantization is based on the ***companding-expanding principle***. According to this principle,
- at the quantization stage, initial non-quantized signals are first nonlinearly transformed by a companding transformation and then the resulting signals are uniformly quantized;
- at the stage of signal restoration, uniform quantization levels are nonlinearly corrected using a corresponding nonlinear expanding transformation.

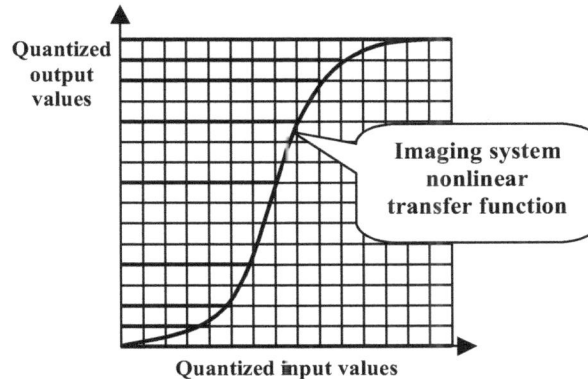

**Fig. 9.23.** Nonlinear signal transformation and the phenomenon of slinging quantization levels together. Thin lines on the figure indicate 16 uniform quantization intervals. The curve shows transfer function of a nonlinear transformation. Bold lines indicate 9 quantization levels left after the nonlinear transformation of 16 quantization levels of input values

Consider now a diagram shown in Fig. 9.24 that explains the principle of correcting signal nonlinear distortions in the described nonlinear quantization scheme.

Let, for the beginning, the nonlinear imaging system transfer function to be corrected (upper right quadrant of the diagram) as well as nonlinear compressing and expanding functions used for signal non-uniform compressor-expander quantization (upper left and bottom right quadrant of the diagram) be known. Project, as it is indicated by arrows in upper left and upper right quadrants of the diagram in Fig. 9.13, boundaries of quantization intervals of the uniform quantizer (left horizontal coordinate axis) at the output of the compressing transformation to its input (upper vertical coordinate axis) and then further to the input of the nonlinear transformation to be corrected (right horizontal coordinate axis). This provides boundaries of the quantization intervals of the non-quantized signal that has to be restored from the distorted and quantized input signal. Within these boundaries, find now optimal quantized representative values of the perfect signal, which minimize the quantization error (dots on the right horizontal coordinate axis). Finally, project the found values back to the input scale of the expanding transformation as indicated by arrows in the bottom right quadrant of the diagram. Obtained values determine the needed corrected output values of the correcting look up table (shown by bold circles in the diagram) that have to be assigned to quantized values of the distorted signal.

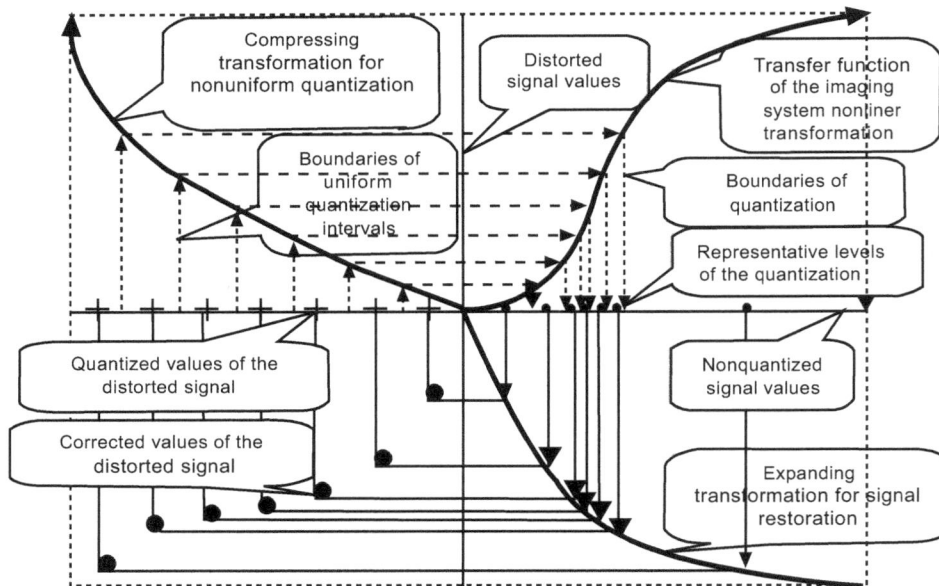

**Fig 9.24.** Principle of correcting nonlinear distortions with an account for effects of quantization

In applications, it very frequently happens that the nonlinear signal distortion transfer function that has to be corrected is not known. In some cases it can be determined directly from the analysis of the distorted signal. A typical example to illustrate such an opportunity is correcting nonlinear distortions of photographic recording of interferograms.

Perfect interferogram signal is a phase modulated sinusoidal signal. If, as it usually happens, the interferogram to be corrected contains many periods of the sinusoidal signal, distribution function of its phase evaluated over the entire recorded interferogram can be regarded uniform. The distribution function of a sinusoidal signal with uniformly distributed phase is known. Therefore, the correcting nonlinear transformation can be found as a transformation that converts the histogram of the distorted interferogram into the histogram defined by the distribution function of the sinusoidal signal with uniformly distributed phase. We will refer to such a transformation as *histogram matching*. Fig. 9.25 illustrates an example of the histogram matching for correcting nonlinear distortions of an interferogram.

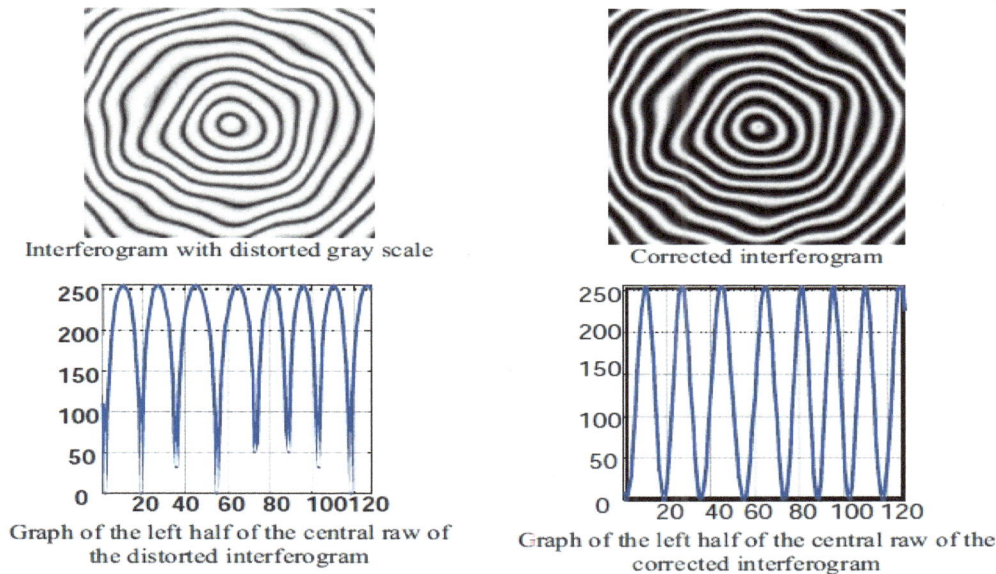

Interferogram with distorted gray scale

Corrected interferogram

Graph of the left half of the central raw of the distorted interferogram

Graph of the left half of the central raw of the corrected interferogram

**Fig. 9.25.** An example of correcting unknown nonlinear distortion of an interferogram

Algorithmically, the histogram matching can be implemented through the ***histogram equalization***, a nonlinear transformation that converts images with whatever gray level histograms into images with a uniform histogram ([1]). For the histogram matching, it is sufficient to find a histogram equalization transformation for the image to be transformed and that for the image with the target histogram. The required histogram matching transformation can then be obtained by combining histogram equalization transformation for the image to be transformed and the transformation inverse to the histogram equalization transformation for the target histogram.

## 9.4   Image resampling

When working with sampled images in computers, one frequently needs to return back to their continuous originals. Typical applications that require building continuous image models are image resampling, image restoration from sparse samples, scaling of results of numerical reconstruction of holograms recorded on different wavelengths. Many different signal resampling and interpolation methods are known in image processing. Most of them, however, are metrologically non perfect and introduce interpolation errors that might be non tolerable in digital holography applications. One of such demanding application, interpolation of mathematical hologram samples for hologram encoding, was already mentioned in Ch. 6. In this section, we describe the ***discrete sinc-interpolation*** that can be regarded the "gold standard" for interpolation of sampled data and compare it with other less accurate methods. More detailed treatment of this topic can be found in Refs. 1 and 6.

### 9.4.1   *Discrete sinc-interpolation: a gold standard of image resampling*

Image resampling assumes reconstruction of a continuous approximation of the original non-sampled image by means of interpolation of available image samples to obtain samples in-between the given ones. For the purposes of the design of the prefect resampling filter, one can regard signal co-ordinate shift as a general resampling operation. This is justified by the fact, that samples of the resampled signal for any arbitrary signal resampling grid can be obtained one by one through the corresponding shifts of the original signal to the given sample position.

Signal resampling as a linear signal transformation can be fully characterized by its point spread function (PSF) or, correspondingly, by its overall frequency response. The optimal shifting re-sampling filter is the filter that generates a shifted copy of the input signal with preservation of the analog signal spectrum in its base band defined by the signal sampling rate and by the number of available signal samples. According to this definition, overall continuous frequency response $H^{(intp)}(p)$ of the optimal $\delta\tilde{x}$-shifting filter (Fourier Transform of its PSF) for the coordinate shift $\delta\tilde{x}$ is, by virtue of the Fourier transform shift theorem,

$$H^{(intp)}(p) = \exp(i2\pi p\delta\tilde{x}) \tag{9.4.1}$$

One can show (see [1, 6]) that discrete frequency response coefficients $\left\{\eta_{r,opt}^{(intp)}(\delta\tilde{x})\right\}$ of the optimal $\delta\tilde{x}$-shift re-sampling filter implemented with Discrete Fourier Transform and applied to discrete signals of $N$ samples must be taken as samples, in sampling points $\left\{r/N\Delta x\right\}$, of its continuous frequency response. Thus, for an odd number of signal samples $N$, coefficients $\left\{\eta_{r,opt}^{(intp)}(\delta\tilde{x})\right\}$ must be set to

$$\eta_{r,opt}^{(intp)}(\delta\tilde{x}) = \frac{1}{\sqrt{N}}\exp\left(i2\pi\frac{r\delta\tilde{x}}{N\Delta x}\right), \quad r = 0,1,...,(N-1)/2$$
$$\eta_{r,opt}^{(intp)}(\delta\tilde{x}) = \eta_{N-r,opt}^{*(intp)}(\delta\tilde{x}) \qquad , r = (N+1)/2,...,N-1 \tag{9.4.2}$$

The second of these two equations is obtained from the requirement that discrete point spread function of the filter must be a sequence of real numbers.

For an even number of signal samples $N$, from the same requirement it follows that the coefficient $\eta_{N/2,opt}^{(intp)}(\delta\tilde{x})$, which corresponds to the signal higher frequency in its base band, must be a real number. Because of that, for even $N$ this coefficient needs a special treatment. A natural setting is assignment:

$$\eta^{(intp)}_{r,opt}(\delta\tilde{x}) = \begin{cases} \exp\left(i2\pi\dfrac{r\delta\tilde{x}}{N\Delta x}\right), & r = 0,1,...,N/2-1 \\[2ex] A\cos\left(\pi\dfrac{\delta\tilde{x}}{\Delta x}\right), & r = N/2 \end{cases} ,$$　(9.4.3)

$$\eta^{(intp)}_{r,opt}(\delta\tilde{x}) = \left(\eta^{(intp)}_{N-r,opt}(\delta\tilde{x})\right)^*, \quad r = N/2+1,...,N-1$$

where $A$ is a weight coefficient that defines signal spectrum shaping at its highest frequency component. In what follows we will consider, for even $N$, the following three options for $A$:

$$\text{Case\_0:} \quad A = 0,$$
$$\text{Case\_1:} \quad A = 1$$
$$\text{Case\_2,} \quad A = 2.$$

One can show (see [1, 6]) that, for odd $N$, point spread function of the optimal resampling filter defined by Eq. 9.4.2 is:

$$h^{(intp)}_n(\delta\tilde{x}) = \text{sincd}\left\{N, \pi\left[n - (N-1)/2 - \delta\tilde{x}/\Delta x\right]\right\}.$$　(9.4.4, a)

where

$$\text{sincd}(N; M; x) = \frac{\sin(x)}{N\sin(x/N)}$$　(9.4.5, a)

is the ***discrete sinc-function*** already introduced in Ch. 3, Sect.3.2, Eq. 3.2.14. For even $N$, Case\_0 and Case\_2, optimal resampling point spread functions are

$$h^{(intp0)}_n(\delta\tilde{x}) = \overline{\text{sincd}}\left\{N; N-1; \pi\left[n - (N-1)/2 - \delta\tilde{x}/\Delta x\right]\right\}$$　(9.4.4, b)

and

$$h^{(intp2)}_n(\delta\tilde{x}) = \overline{\text{sincd}}\left\{N; N+1; \pi\left[n - (N-1)/2 - \delta\tilde{x}/\Delta x\right]\right\},$$　(9.4.4, c)

correspondingly, where a ***modified sincd-function*** $\overline{\text{sincd}}$ is defined as

$$\overline{\text{sincd}}(N; M; x) = \frac{\sin(Mx/N)}{N\sin(x/N)}$$　(9.4.5, b)

One can easily see that Case\_1 is just the average of Case\_0 and Case\_2:

$$h^{(intp2)}_n(\delta\tilde{x}) = \left[h^{(intp0)}_n(\delta\tilde{x}) + h^{(intp2)}_n(\delta\tilde{x})\right]/2 = \overline{\text{sincd}}(\pm1; N; x) =$$
$$\left[\overline{\text{sincd}}(N-1; N; x) + \overline{\text{sincd}}(N+1; N; x)\right]/2.$$　(9.4.4, d)

Filter defined by Eqs. 9.4.4 is the $\delta\tilde{x}$ - ***fractional shift*** filter that preserves samples of continuous signal spectrum in its base band. All resampling filters with other discrete frequency response, or, correspondingly, point spread function, will distort signal spectrum and therefore introduce interpolation error. It is in this sense that discrete sinc-interpolation can be regarded the "gold standard" of interpolation of sampled data.

An important application issue of the discrete sinc-interpolation is its computational complexity. The described $\delta\tilde{x}$-fractional shift filter that implements discrete sinc-interpolation is designed in DFT domain. Therefore it can be straightforwardly implemented using Fast Fourier Transform with the computational complexity of $O(\log N)$ operations per output signal sample, which makes it competitive with other less accurate interpolation methods ([6]).

From the application point of view, the only drawback of such an implementation is that it tends to produce signal oscillations due to the boundary effects caused by the circular periodicity of convolution implemented in DFT domain. These oscillation artifacts can be virtually completely eliminated, if the discrete sinc-interpolation is implemented in DCT domain (for details see [6]).

### 9.4.2   *Optimal re-sampling filter: discussion and interpretation.*

Discrete sinc- functions sincd and $\overline{\text{sincd}}$ given by Eqs. 3.2.14 and 9.4.5 define point spread functions of the ideal digital low-pass filter, whose discrete frequency response is a rectangular function. Discrete sinc-functions are discrete analogs of the continuous sinc-function. However, these functions must not be confused.

Continuous sinc-function is an a-periodic function with an infinite support. It is a convolution kernel that preserves spectra of continuous signals in their base band defined by the sampling rate. For band-limited signals, it provides signal perfect reconstruction from their samples provided infinite number of the samples is available. For signals that are not band-limited, it provides least mean square error signal reconstruction under the same condition of the availability of infinite number of signal samples.

Continuous sinc-function cannot be implemented in reality due to its infinite support. As a practical substitute for sinc-function, a windowed sinc-function is considered frequently for interpolation purposes in the belief that this is the best one can do in the discrete case with finite number of signal samples. However, the windowed sinc-function as a convolutional interpolation kernel deviates from the discrete sinc-function and therefore it provides worse interpolation accuracy compared to that provided by the discrete sinc-interpolation.

Discrete sinc-interpolation is a periodical function of sample indices. One can evaluate similarity and dissimilarity of continuous and discrete sinc-functions using plots shown in Fig. 4.3, Ch. 4. Discrete sinc-function secures perfect, for the given number of signal samples, discrete signal re-sampling with preservation of the corresponding continuous signal spectra in their sampling points. Continuous frequency response of discrete-sinc interpolators coincides with that of continuous sinc-interpolator in spectral sampling points defined by the number of the discrete signal samples, but may deviate from it between sampling points, deviations being dependent on discrete signal boundary conditions.

In order to illustrate perfect interpolation capability of the discrete sinc-interpolation in comparison to other numerical interpolation methods we will provide experimental data of comparing interpolation methods in image rotation application adopted from Ref. 6. Test images were subjected in experiments to multiple rotations so as the total rotation angle be multiple of $360°$. Then the rotation error was evaluated as a difference between initial and rotated images.

As test images, two images were used: an image of a printed text ("Text"-image, Fig. 9.26, a) and a computer generated pseudo-random image with uniform DFT spectrum ("PRUS"-image, Fig. 9.26,b). The "Text" image was used to evaluate readability of the text after image rotation for different rotation method. The "PRUS" image was used for demonstrating distortions of signal spectra caused by the re-sampling.

In these experiments, nearest neighbor, bilinear and bicubic interpolation Matlab programs **imrotate.m** and **interp2.m** from Image Processing Toolbox were used. For discrete sinc-interpolated rotation, a Matlab code was written using standard Matlab tools that implements the 3-step rotation algorithm ([6]) through the above-described DFT-based fractional shift filter.

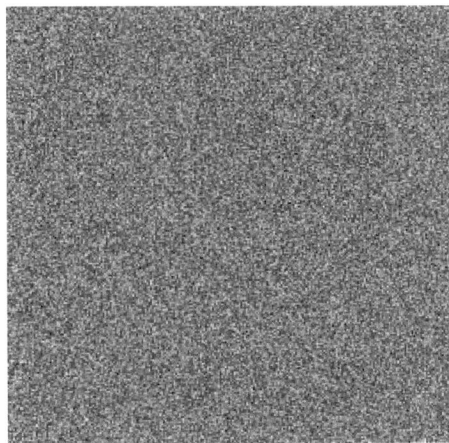

a)                                         b)

**Fig. 9.26.** Test images used in the experiments: (a) – "Text"-image; b) –" PRUS"-image

Results of comparison, for rotation of the test image "Text", of three standard methods of image re-sampling, nearest neighbor, bilinear and bicubic ones, and that of discrete-sinc interpolation are shown in Figs. 9.27, a)-d). Images shown in the figure clearly demonstrate that, after 60 rotations though $18°$ each, standard interpolation methods completely destroy the readability of the text, while discrete sinc-interpolated rotated image is virtually not distinguishable from the original one.

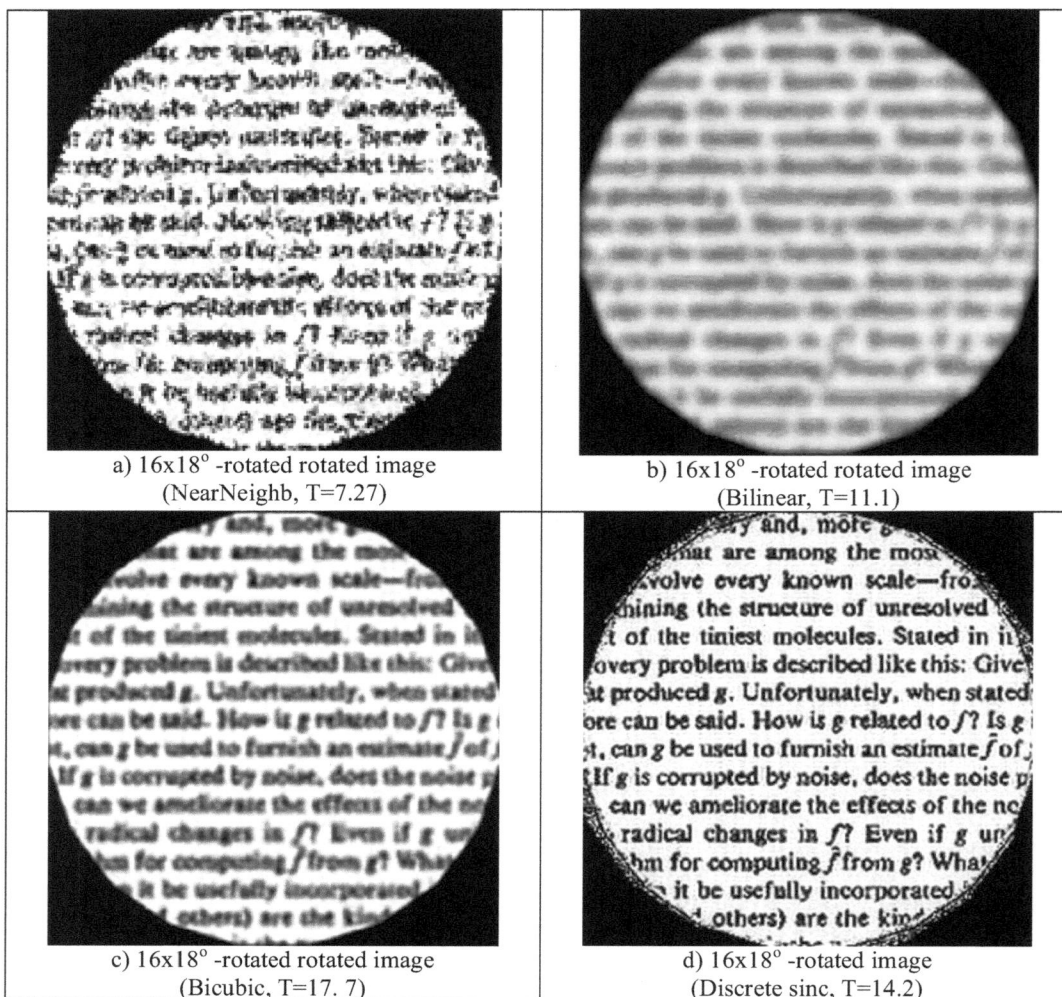

| a) $16 \times 18°$-rotated rotated image (NearNeighb, T=7.27) | b) $16 \times 18°$-rotated rotated image (Bilinear, T=11.1) |
| c) $16 \times 18°$-rotated rotated image (Bicubic, T=17. 7) | d) $16 \times 18°$-rotated image (Discrete sinc, T=14.2) |

**Fig. 9.27.** Discrete sinc-interpolation versus conventional interpolation methods: results of multiple rotation of the test image "Text": a) - nearest neighbor interpolation; b) bilinear interpolation; c) – bicubic interpolation; d) – discrete sinc-interpolation. Numbers in brackets indicate processing time

Nearest neighbor, bilinear and bicubic interpolations are spline interpolations of the first, second and third order. The higher the spline order the higher interpolation accuracy and the higher their computational complexity. However, higher order spline interpolators still underperform the discrete sinc-interpolation (corresponding comparison data one can find in [6]).

In the analysis of the interpolation errors, it is very instructive to compare their power spectra. Fig. 9.28 shows rotation error spectra for $10 \times 36°$ rotations, using bicubic, higher order spline ("spline351", [7]) and the discrete sinc-interpolation, of PRUS-image that was low pass pre-filtered to 0.7 of the base band in horizontal and vertical directions. As a result of such a low pass filtering, aliasing error components are excluded, and one can see from the figure that discrete sinc-interpolation, as opposite to other methods, does not produce, in this case, rotation errors at all (few components of noise spectrum still visible in Fig. 9.28, c) are intentionally left after the pre-filtering as reference points that keep, for display purposes, same dynamic range of error spectra for all compared methods).

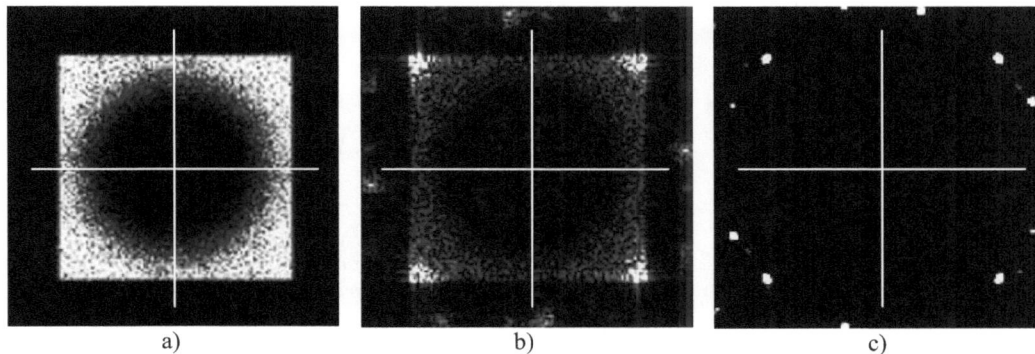

a)             b)             c)

**Fig. 9.28.** Rotation error spectra for 10X36° rotations of PRUS image low pass filtered to 0.7 of its base band (bright square in fig. a)) for bicubic (a), Mems(5,3,1) (b) and discrete sinc-interpolation (c) methods. All spectra are shown in frequency coordinates (white lines) centered at spectrum zero frequency (dc-component), image lightness being proportional to error spectra intensity; bright points in fig. c) are test image spectral high frequency components at corners of the base band intentionally left for comparison purposes; they appear due to aliasing on the borders of the spectral domain in course of rotation

# References

1. L. Yaroslavsky, Digital Holography and Digital Image Processing, Kluwer Academic Publishers, Boston, 2004
2. L. Yaroslavsky, A. Shefler, Statistical characterization of speckle noise in coherent imaging systems, in: Optical Measurement Systems for Industrial Inspection III, SPIE's Int. Symposium on Optical Metrology, 23-25 June 2003, Munich, Germany, W. Osten, K. Creath, M. Kujawinska, Eds., SPIE Proceedings Series , v. 5144, pp. 175-182
3. L. Yaroslavsky, Space Variant and Adaptive Transform Domain Image and Video Restoration Methods, In: Advances in Signal transforms: Theory and Applications, J. Astola, L. Yaroslavsky, Eds. , EURASIP Book Series on Signal Processing and Communications, Hindawi, 2007
4. D.L. Donoho and I.M. Johnstone, Ideal Spatial Adaptation by Wavelet Shrinkage, Biometrica, 81(3): 425-455, 1994
5. L. Yaroslavsky, N. Merzlyakov, Methods of Digital Holography, Consultant Bureau, N.Y., 1980
6. L. Yaroslavsky, Fast Discrete Sinc-Interpolation: A Gold Standard for Image Resampling, In: Advances in Signal Transforms: Theory and Applications, J. Astola, L. Yaroslavsky, Eds. , EURASIP Book Series on Signal Processing and Communications, Hindawi, 2007
7. A. Gotchev, "Spline and wavelet based techniques for signal and image processing," Doctoral thesis, Tampere University of Technology, Tampere, Finland, 2003, publication 429.

# Index

www.ingramcontent.com/pod-product-compliance
Lightning Source LLC
Chambersburg PA
CBHW041702210326
41598CB00007B/506

*9781608053155*